Gravitational Waves:
An Overview

Synthesis Lectures on Wave Phenomena in the Physical Sciences

Editor
Sanichiro Yoshida, *Southeastern Louisiana University*

Gravitational Waves: An Overview
David M. Feldbaum

ISBN: 978-3-031-01485-7   paperback
ISBN: 978-3-031-02613-3   ebook
ISBN: 978-3-031-00357-8   hardcover

DOI 10.1007/978-3-031-02613-3

A Publication in the Springer series
SYNTHESIS LECTURES ON WAVE PHENOMENA IN THE PHYSICAL SCIENCES

Lecture #3
Series Editor: Sanichiro Yoshida, *Southeastern Louisiana University*
ISSN pending.

Gravitational Waves: An Overview

David M. Feldbaum
Southeastern Louisiana University

SYNTHESIS LECTURES ON WAVE PHENOMENA IN THE PHYSICAL SCIENCES #3

ABSTRACT

Gravitational wave (GW) research is one of the most rapidly developing subfields in experimental physics today. The theoretical underpinnings of this endeavor trace to the discussions of the "speed of gravity" in the 18th century, but the modern understanding of this phenomena was not realized until the middle of the 20th century. The minuteness of the gravitational force means that the effects associated with GWs are vanishingly small. To detect the GWs produced by the most enormously energetic sources in the universe, humans had to build devices capable of measuring the tiniest amounts of forces and displacements.

This book delves into the exploration of the basics of the theory of GW, their generation, propagation, and detection by various methods. It does not delve into the depths of Einstein's General Relativity, but instead discusses successively closer approximations to the full theory. As a result, the book should be accessible to an ambitious undergraduate student majoring in physics or engineering. It could be read concurrently with standard junior-level textbooks in classical mechanics, and electromagnetic theory.

KEYWORDS

general relativity, quadrupole radiation, laser interferometry, black hole binaries

Contents

Introduction

Gravitational Wave (GW) research is one of the most rapidly developing subfields of today's experimental physics. The theoretical underpinnings of this endeavor trace back before Einstein, to the discussions of the "speed of gravity" by Laplace in the 18th century, and to the considerations of "gravimagnetic" fields by Heaviside in the 19th century. The modern theoretical understanding of GWs started to appear in Einstein's own work, and was gradually developed over the first half of the 20th century. The second half of the 20th century brought the realization that GW may actually be detectable, which resulted in the exploration of the wide variety of their possible sources, as well as the development of various detectors. The first observation of the *effects* of GW radiation happened in 1974, and the first direct detections were being unsuccessfully attempted starting in the 1960s. The design of the detectors which would eventually succeed was published in 1973, they were built starting in the late 1990s, and, after a number of upgrades and developments throughout the early 2000s, achieved spectacular results in 2014. Within the next several decades a number of additional detectors are expected to be built, based both on the same and on completely different principles.

This book goes into the exploration of the basics of the theory of GW, their generation, propagation, and detection by various methods. The book should be accessible to an undergraduate student. It could be read concurrently with standard junior-level textbooks in classical mechanics (such as [4]), electromagnetic theory [2] and general relativity [8].

PART I

Review of Wave Motion

CHAPTER 1

The Wave Equation

1.1 ONE SPATIAL DIMENSION

From one point of view, gravitational waves (GWs) represent one particular example of general wave propagation. Mathematically wave propagation is described by *a wave equation*: a partial differential equation (PDE) in time and space variables, that has propagating solutions. Given initial conditions in the form of a disturbance at time zero, the wave-like solution should show a related disturbance at a different point in space at a later time. The *non-dispersive wave equation*

$$\frac{\partial^2 f}{\partial x^2} = v^2 \frac{\partial^2 f}{\partial t^2} \tag{1.1}$$

is the prototypical simplest case of a wave equation. It possesses propagating disturbances which are unchanged with time:

$$f(x,t) = f(x \pm vt, 0), \tag{1.2}$$

where $f(x,t)$ is any differentiable function.

Many other differential equations possess propagating solutions, and therefore may be characterized as wave equations. These include such well-known equations as:

- transmission line with resistance R, inductance L, capacitance C, and leakage conductance G:

$$\frac{\partial i}{\partial x} + C \frac{\partial v}{\partial t} = -Gv$$

$$\frac{\partial v}{\partial x} + L \frac{\partial v}{\partial t} = -Rv$$

- or the vibrating beam equation

$$\frac{\partial^4 y}{\partial x^4} + \frac{\partial^2 y}{\partial t^2} = 0.$$

Partial differential equations of order two may be classified into three classes (*Elliptical, Parabolic, Hyperbolic*), with the solutions behaving similar to Laplace, Heat, and Wave equation, respectively. The solutions of the hyperbolic partial differential equations (PDEs) have a wavefront which propagates with a finite speed (as opposed to both the parabolic PDEs, which lack sharp wavefront and exhibit infinite speed of propagation, and to the elliptical PDEs, which describe static situations).

All of this is relatively irrelevant for the basics of GWs research. The more complicated wave equations are relevant when waves propagate in various *dispersive media*, which affect the speeds of propagation and/or amplitudes of the waves. The basic wave described by Eq. (1.1) is the type of the wave that occurs when the disturbance propagates with a constant speed and amplitude, unaffected by the medium. This is the type of wave propagation that occurs when an EM wave propagates in vacuum.

What about gravitational waves? Whether or not gravitational fields satisfy the basic wave equation was an active topic of discussion in the first half of the 20th century. General Relativity—Einstein's theory of gravitational fields—is different from the Maxwell's theory of electromagnetism in many major respects. The possibility that gravitational field may take form of a wave propagating in accordance with the basic wave equation has been posed and disproved several times by Einsteins himself, as well as by others.

Presently, it is understood that the weak gravitational fields in vacuum may satisfy the basic wave equation (1.1). All discussion in this book will be centered on these types of GWs.

1.2   WAVES AND OSCILLATIONS

When we think about waves we usually don't just think of a propagating disturbance of an arbitrary shape. The usual picture that comes to mind at the mention of waves is of something moving periodically "up-and-down." Regular oscillations, such as those of a piece of wood bobbing on the surface of a pond come to mind naturally. The consideration of oscillations is also necessarily invoked when the wave motion is scientifically examined.

So how come no mention of oscillatory motion has been made in the previous section? The wave was defined there by Eq. (1.2), which describes any, arbitrary disturbance propagating with time. Where do regular oscillations come from?

At the mathematical level the answer comes from *Fourier analysis*. According to that branch of mathematical physics, a physical disturbance of arbitrary shape may be "decomposed" into *harmonic* waves. Harmonic waves are waves that look like pure sine or cosine curves. They are the simplest to work with. By using Fourier analysis we can add together a sufficient number of carefully chosen sines and cosines, to represent any physically interesting function.

Fourier analysis is an indispensable tool for the study of any branch of physics at the advanced level. At the more intermediate level we usually work with individual harmonic waves, at most adding a few of them.

Mathematically, a harmonic wave may be described in a number of equivalent ways. Fundamentally, to write the expression for a harmonic wave, we need to take the definition of the general wave, $f(x,t) = f(x \pm vt, 0)$, and then to specify that function $f(x,t)$ is just a sinusoid:

$$f(x,t) = A \cos[k(x \pm vt) + \phi_0].$$

The *amplitude A* in front of the sinusoid specifies the actual physical quantity that is propagating, and so relates mathematics to physics. For example, if the waves in question are sound waves, then the amplitude A represents the magnitude of the pressure (or density) deviations, and has the units of pressure (or density). The $\cos[k(x \pm vt) + \phi_0]$ describes the sinusoid whose peaks travel at velocity v. Obviously, not only the peaks, but the troughs, and any other particular *phase* of the wave travels at the same velocity, which is called the *phase velocity*. The distance between two consecutive peaks (or, again troughs, or, better, between any two consecutive points of the same phase) is the *wavelength* $\lambda = 1/k$. An observer sitting at one particular point along the $\hat{x}$-axis sees oscillations in time whose frequency is $\omega = kv$. It is more common to write the expression for the propagating wave using the frequency rather than the phase velocity:

$$f(x, t) = A \cos(kx \pm \omega t + \phi_0).$$

Finally, the free phase ϕ_0 determines whether the wave "started out" with a peak, or in a trough, or somewhere in between.

A mathematically equivalent, but more convenient form of describing sinusoid signals uses *Euler's formula*:

$$e^{ix} = \cos x + \iota \sin x, \tag{1.3}$$

where ι is the *imaginary unit*. Using this Euler's formula, any sinusoid may be expressed as the real part of complex exponent. In particular,

$$\cos(kx \pm \omega t) = \text{Real}\{e^{\iota(kx \pm \omega t)}\} \tag{1.4}$$

and so the harmonic wave may be written as

$$f(x, t) = \text{Real}(A e^{\iota(x \pm vt)}). \tag{1.5}$$

1.3 THREE SPATIAL DIMENSIONS

The qualitative behavior of the solutions of elliptical (e.g., Laplace) and parabolic (e.g., heat) equation does not change with the increase in the number of dimensions. This is not the case for the hyperbolic (wave) equations. A variety of possible solutions exist in the three-dimensional case. Only one particular type of solution (the *plane wave* solution) preserves the property of propagating without change:

$$f(\mathbf{r}, t) = f(\mathbf{r} \cdot \hat{n} - vt, 0). \tag{1.6}$$

A variety of other solutions are constructed for various cases, the most important one being the *spherical wave* solutions:

$$f(r, t) = \frac{f(r - vt)}{r} \tag{1.7}$$

which represents a disturbance which propagates radially outward from the origin. The amplitude of this wave decreases as it propagates away from the origin.

A special case of the spherical wave is the harmonic spherical wave, which we will write right away using the exponential notation:

$$f(r,t) = \frac{A}{r} e^{ik(r-vt)}. \tag{1.8}$$

The amplitude of this wave is not A, but A/r, and the quantity A is the *source strength*. Just as in case of a wave in one dimension, the quantity A carries the information about the actual quantity being carried as a wave.

It is obvious that such a spherically symmetric solution should be extremely important. Many of the actual sources of waves in the physical universe are point-like or almost point-like. For instance, on the basic level the sources of light are individual atoms or molecules, whose sizes are thousands times smaller than the wavelengths of the emitted light. Even though in most classical situations these point-like sources combine to produce cumulative effects, the understanding of the spherical waves produced by point-like sources is necessary for understanding of the cumulative effects.

Such considerations are even more relevant in the case of GWs. It might seem that the sources of GWs are not at all point-like: they are superdense astrophysical objects. However, for one, superdense does not at all mean superlarge: the diameters of the typical black holes and neutron stars count in tens (and not in tens of millions) kilometers. More relevant, however, is that these sources are located at astronomical distances from us: just as stars and even entire galaxies appear as point light sources to the terrestrial observers, so do astronomical sources of GWs.

So the understanding of spherical waves produced by point sources is of the utmost importance for a student of GWs. It is important, however, to realize right away that the simplest harmonic spherical wave of Eq. (1.8) cannot, actually, be a valid solution either for EM waves or for GWs. There is an infinite number of more complicated possible spherical wave forms. They are similar in that they all expand outward, and their amplitudes all decrease proportionally with the distance from the source. Unlike the simplest case above, however, the amplitudes and phases of these waves depend not only on the distance, but also on the direction from the origin. While the radiation pattern of Eq. (1.8) is spherically symmetric, the radiation patterns of these more complicated spherical waves have "lobes."

The next simplest spherical wave is a wave produced by the *dipole radiation* with two lobes in the opposite directions. The dipole radiation is typical for EM waves. The GWs do not produce either uniform *or* dipole radiation patterns. They may only be produced by *quadrupole radiation*, with, yet, a more complicated radiation pattern: four lobes in a "+" or "x" pattern. EM waves also can propagate in this pattern, but dipole radiation, if it can exist, dominates quadrupole radiation.

Either EM or GW can propagate in even more complicated patterns with 2^n lobes, where n is an arbitrary integer. However these radiation modes all have smaller radiation strength and are often ignored.

CHAPTER 2

Waves of What?

Classical physics contains several prominent separation lines, which have become erased, or, at least, blurred in modern physics: matter versus energy, past versus present versus future, even existence versus non-existence.

The line separating waves from particles is one the most famous for being erased in the early 20th century. In quantum mechanics, any particle has a wave function associated with it, while any wave may be regarded as carrying an indeterminate number of quanta. Confusion sometimes arises where the students of physics imagine a wave that "carries" particles, not unlike an ocean wave "carrying" a surfer. Nothing of the kind actually happens. The quantum-mechanical particles are the quanta of energy of a quantum mechanical wave. The *massless* particles, such as photons, should not even be imagined as discrete particles: they cannot be localized. For example, a photon in an interferometer cannot be said to be located at a particular point in space between a pair of mirrors—in fact, it cannot even be said to be located in one of the two (or more) arms of an interferometer.

The same applies to the GWs. Although there is, as yet, no quantum theory of gravity, most physicists believe that *gravitons*—quanta of gravitational fields—exist. Assuming they do, they are massless, and, therefore, cannot be localized in space.

At any rate, we will not be going further into the quantum-mechanical details. Most of the knowledge necessary to understand the phenomenon of GWs lies outside of quantum physics.

So then, from a non-quantum point of view, waves of what are the GWs? The aforementioned ocean waves are the waves of the elevation of the local surface above and below the average level. The waves on a string are the waves of the displacement of the string from the undisturbed position. Sound waves are the waves of compression and decompression of the medium in which the sound is traveling, that is the waves of pressure and density.

In each of the above situations, not one but two statements have actually been made. One statement is about the *medium* of propagation (the surface of the ocean, string, air). The second statement is about the nature of the propagating quantity (the elevation, displacement, pressure and density).

The reader, most likely, knows that until the early 20th century physicists assumed that the same two-part answer should be given to the question of "waves of what are the EM waves?" It was postulated that there existed a medium, "luminiferous aether," or "light-carrying ether," which permeated all space and matter, but did not interact with any matter. The only reason for, and evidence of its existence was the propagation of EM waves.

Fizeau experiment and *Michelson–Morley experiment* conducted in the 19th century demonstrated how special the properties of this substance would have to be. Finally, in 1905, Einstein's Special Relativity eliminated ether from physics altogether.

Relativity's answer to the first question (what is the medium of propagation) is that either the question does not need to be asked at all, or that the answer is very simple: the EM waves propagate in a vacuum. The second question (what is propagating), of course, has the definitive answer: the electric and magnetic fields are the propagating quantities. These fields are detectable by their interactions with charged particles and electric currents.

What about GWs? Just like the EM waves, GWs propagate in vacuum at the speed of light. Should we say that the propagating quantity is the gravitational field? Is there a quantity analogous to the magnetic fields?

The answers to both of the above questions are "yes," but that does not tell the whole story. It is possible to treat gravitational fields in a manner analogous to the electric fields and to obtain *some* correct answers in certain situations, up to a point. The fields analogous to the magnetic fields, exist in the case of gravitation as well. However, the full theory of gravitation is much more involved than the theory of electromagnism. In a sense, there are *more* fields associated with the gravity, moreover, there are more *sources* of these fields than just the masses and their motions.

These subjects will be explored to some modest extent in the upcoming chapters. But before going into those details we should look at a simpler version of what constitutes GWs.

General Relativity (GR) is the *geometrical* theory of gravity. The notion of gravitational force is abandoned, and the gravitational attraction is shown to result from force-free motion within the curved spacetime. The fact that an object released close to the Earth's surface with no initial velocity then proceeds to approach the Earth is explained as a combination of two facts of relativity:

(1) From the point of spacetime (as opposed to just space), every object is always traveling at the speed of light. Any object possesses a *4-velocity* vector U

$$U_\mu = \frac{dx_\mu}{d\tau} \tag{2.1}$$

whose spacetime magnitude equals c, the speed of light in the vacuum. From the point of view of the spacetime, an object that is stationary in the three spatial dimensions (in a particular reference system) just has its velocity 4-vector aimed in the time-direction.

(2) Any mass distorts spacetime. The distortion is such that nearer the masses the space-time intervals contract.

The motion of a particle in the curved spacetime looks like a motion with a constant acceleration in the 3-dimensional (3-D) space. A common misunderstanding should be be corrected right away. A popular image that is meant to help in visualization of GR is that of a rubber sheet deformed by a heavy mass. A smaller ball is then shown traveling upon the curved surface. Such an illustration is, actually, very wrong. The familiar free-fall acceleration on the Earth's surface has nothing to do with the deformation of space. Instead, the effects of gravitational fields upon slowly moving objects are entirely due to the deformations in the time scale. The time "ticks slower" closer to the gravitating mass, which leads to the alteration of path of nearby objects.

The distortion of the space part of spacetime leads to the effects that are much more difficult to detect under normal (for humans) circumstances. These effects become more significant for fast moving objects. In particular, for light the effects of spatial curvature become comparable to the effects of time distortion. One of the first experimental confirmations of GR came from the observations of the deflection of the starlight by the gravitational pull of the Sun. The Newtonian mechanics-based calculations of the deflection of light by large masses had been made several times since the late 18th century. These calculations predict only one-half of the experimentally observable deviation. GR predictions, on the other hand, agree with the experiment.

Other experimentally observable effects of spatial curvature require either long-time observations, or very high-precision measurements. The measurement of GWs is one of the most dramatic such measurements to date.

GWs are the propagating distortions in the spacetime curvature. In fact, it is, mostly, the spatial part of spacetime that is being distorted in a GW. These distortions are minute and they typically do not spend much time time traveling through the measuring devices. Therefore, they require extremely precise measurements to become detectable.

2.0.1 POLARIZATION: THE SHAPE OF A WAVE

Now that we said that that GWs are the propagating distortions of spacetime, we would like to try to imagine what they actually look like. Almost everyone has some visual experience with the waves on a string or waves on the water surface. We can easily relate to the fact that an ocean wave mostly distorts the water surface in the vertical direction (although water waves are very complicated phenomena, in particular, water actually moves in vertically-oriented circles within the waves). We can likewise easily realize that a horizontally stretched string may vibrate both horizontally and vertically.

Other waves are not as commonly seen, but may be readily imagined. A sound pressure-density wave is easily visualized. It may also be modeled by a longitudinal wave in a spring (Fig. 2.1).

The EM waves are, certainly, more difficult to visualize directly. J. C. Maxwell, reportedly, had to imagine space filled with mechanical gears to visualize electromagnetic fields. Without taking up such obvious crutches, one can visualize EM waves within dielectric media. Such a wave separates the positive and negative charges which constitute the material, producing a wave of polarization within the material.

Before the 20th century, top scientists postulated the existence of *luminiferous aether*: an invisible substance which permeates both the empty space and physical objects, and which almost does not interact with any physical objects. The sole purpose of the aether was to be the carrier medium for the EM waves. The imagined existence of ether provided a means of vi-

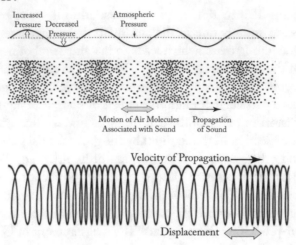

Figure 2.1: Density waves.

sualization: one could imagine the light wave propagating within the aether just as one would imagine it propagate within a crystal.

The idea of aether was "banished" by Special Relativity in the early 20th century. Since then physicists have to use abstraction. e rely on either of the two closely related means of visualization. The first is practical: we imagine the effect of a wave (or, generally, any field) on something observable. In the case of electromagnetic fields we may imagine the effect of a passing wave upon a charged particle, either free or a bound one, or upon a current loop. A passing EM wave will oscillate a charge particle.

The second means of visualization is abstract. We draw vector fields to represent the directions of the fields at a certain moment in time. There are several ways of drawing vector fields, such as drawing field lines and drawing arrows. All of these methods are, in some way, problematic [1], but, used with care, they can provide useful intuition about abstract concept.

How should we visualize GWs, and gravitational fields in general? Should we be imagining vectors of gravitational attraction, in analogy with the vectors of electric fields?

Although there will be made much of the analogy between the electromagnetic and gravitational fields in the upcoming chapters, it should be realized at the outset that gravitational fields are very different from the electromagnetic ones. On the major differences is that **the gravitational fields are not vector fields**. This may sound very confusing since the gravitational attraction is the first most basic example of a directional force that every earthbound human encounters in infancy. Newtonian law of universal gravitation is also a vector law, analogous to the Gauss's law in electrostatic.

Nonetheless, gravitational fields are not vector fields. The proper way of imagining them is not through their action upon a single massive particle. Gravitational fields, in a dramatic de-

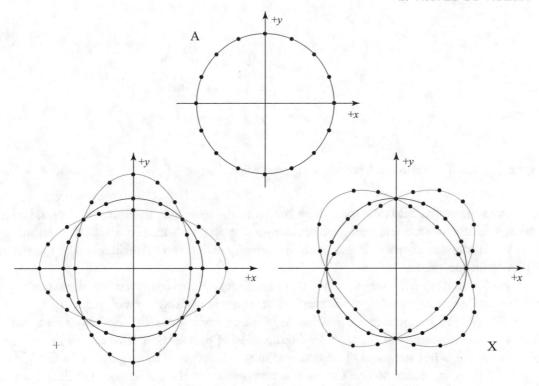

Figure 2.2: An illustration of an effect of a gravitational wave upon small spherical volume. The strain required to produce the distortions shown in this figure is approximately $h = 1.3$. The strain that actually has to be measure by detectors is less than $h = 10^{-19}$.

parture from the electric fields, cannot be measured by their action upon single isolated particles. This fact is directly related to the central starting principle on which the entire edifice of the GR rests: the equivalence principle.

Mathematically, a GW may be represented by a *second rank tensor*. This is a more complicated object than a vector (in fact, an ordinary vector is a *first rank tensor*). The word "tensor" is derived from the word "tension," and hints at the kinds of situations they started out describing. The tensions in a brick or in a pipe (or in anything else that a mechanical engineer might find interesting) are not representable just by vectors. Under external loads, any small portion of the said brick may be compressed, stretched, rotated, and shifted in any direction. This requires more than just three components that a vector has.

To imagine imagine such distortions, we may visualize how small spherical volumes (rather than just individual points) inside the brick change. A single such sphere may get displaced from its original position, *and* get stretched in some directions and squeezed in others. In general, the sphere will become an ellipsoid, with three different axes (Fig. 2.2).

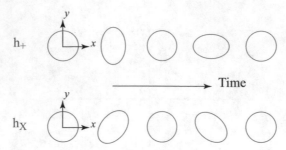

Figure 2.3: An illustration of a propagating gravitational wave.

A gravitational field does something similar to the spacetime itself: it stretches and compresses it. Unfortunately, imagining small volumes of spacetime is more difficult, than just imagining small volumes of space. Physicists have come up with several visualizations of the effects of gravity on spacetime.

Fortunately, in the case of a GW, there are fewer options for spacetime distortions, and we may get by by just thinking of the small volumes of space changing with time.

First, a GW is a *transverse wave*. Just as in the case of EM waves (or of waves on a string) there are *no* components of the wave in the direction of propagation. This means that a small sphere of space is not stretched or compressed in the direction of propagation of a GW. If we set up the coordinate axes so that the wave is propagating in the $\hat{z}$ direction, then the space in the $\hat{z}$ direction is not affected by the wave.

Second, in the directions perpendicular to the direction of propagation the amount of stretching-compression is balanced. The stretching in, say, direction $\hat{x}$ is exactly balanced by the equal amount of compression in direction $\hat{y}$.

So this is how we should imagine a GW: a disturbance, which distorts the space in two perpendicular directions at the same time (Fig. 2.3).

CHAPTER 3

Plane Gravitational Waves

Before we start diving into the fundamental theory trying to understand the origins of GWs, let us explore their properties in the simplest case: that of a plane GW. The reader should already have a "picture" of what one looks like from the previous chapter. Now we are going to put it on a more mathematical basis.

Gravitational fields distorts the spacetime. To understand what that means mathematically we should examine spacetime geometry.

The geometry of spacetime is describable by the *metric tensor*. Metric tensor contains a set of numbers that have to be inserted into Pythagorean theorem to make it work for distorted spacetimes.

On a flat sheet of paper Pythagorean theorem is (hopefully) known from elementary mathematics. If the sides of a right triangle are a and b, and the hypotenuse is c, then $c^2 = a^2 + b^2$. For most purposes in physics the value of Pythagorean theorem lies in the fact that the sides of the triangle can be made into the coordinate axes. Then the length of some vector in that coordinate system may be found through the use of Pythagorean theorem: $v^2 = x^2 + y^2$.

On a curved surface, for example on the surface of a sphere, Pythagorean theorem requires modification. First, it is convenient to talk only about infinitesimally small right triangle. The lengths of the sides of such a right triangle are dx and dy. The generalized Pythagorean theorem for an infinitesimal right triangle drawn on a curved surface takes on a form

$$dv^2 = g_{xx}dx^2 + g_{yy}dy^2 + 2g_{xy}dxdy, \tag{3.1}$$

where g_{11}, g_{22}, and g_{12} are just some real numbers, which show how curved the surface is. These numbers, collectively, are the *metric tensor* of the surface.

It is convenient to arrange these numbers in a matrix:

$$g_{xy} = \begin{bmatrix} g_{xx} & g_{xy} \\ g_{yx} & g_{yy} \end{bmatrix}. \tag{3.2}$$

Matrices are usually introduced through their usefulness in dealing with linear system, in particular, with systems of linear equations. However homogeneous second degree polynomials in any number of variables are also conveniently representable by matrices. For example, to represent the generalized Pythagorean theorem in Eq. (3.1) above, we may write

$$dv^2 = [dx \ dy] \begin{bmatrix} g_{xx} & g_{xy} \\ g_{yx} & g_{yy} \end{bmatrix} \begin{bmatrix} dx \\ dy \end{bmatrix}. \tag{3.3}$$

Homogeneous second degree polynomials, or *quadratic forms*, are commonly encountered in physics.

So, the geometry of an ordinary flat Euclidean plane is described using Pythagorean theorem:

$$dv^2 = [dx \ dy] \begin{bmatrix} 1 & 0 \\ 0 & 1 \end{bmatrix} \begin{bmatrix} dx \\ dy \end{bmatrix}.$$

So the metric tensor of a flat plane is just a unit 2×2 matrix. What if a different set of numbers is used? What if we take, for example,

$$\begin{bmatrix} 1 & 0 \\ 0 & 3 \end{bmatrix}$$

to be the metric tensor? The modified Pythagorean theorem then says $dv^2 = dx^2 + 3dy^2$. Does this describe a distorted, curved surface?

Some thinking should convince the reader that the surface described by such new metric tensor is still the old flat sheet. The only difference is that now the distances along the x-axis are marked with a different ruler than the distances along the y-axis.

3.1   EXAMPLES OF TRIVIAL METRIC TENSORS

3.1.1   CHANGE OF ONE AXIS

To clarify the claim made at the end of the previous section, suppose we decided, for some strange reason, to use different units for measuring the distanced along the y-axis direction than along the x-axis. For example, our units of "length" in the x-directions are meters (m), while along the y-direction we only use centimeters (cm). How can we calculate the distance between two arbitrary points on the plane, say, points whose coordinates are [$x = 2$ m; $y = 200$ cm], and [$x = 3$ m; $y = 300$ cm]?

Let us suppose that we want to obtain the answer in meters. One obvious way is to convert the measurements along the x-axis to meters, then use the basic Pythagorean theorem. We can, however, obtain the result by using a modified Pythagorean theorem, by, effectively, placing the unit conversions inside the metric tensor. The metric tensor for this situation is simply

$$\begin{bmatrix} 1 & 0 \\ 0 & (0.01 \text{ m/cm})^2 \end{bmatrix},$$

and the calculation of the distance becomes

$$\Delta d^2 = [\Delta x \ \ \Delta y]\begin{bmatrix}1 & 0 \\ 0 & (0.01\text{m/cm})^2\end{bmatrix}\begin{bmatrix}\Delta x \\ \Delta y\end{bmatrix}$$

$$= [3\text{ m} - 2\text{ m} \ \ 300\text{ cm} - 200\text{ cm}]\begin{bmatrix}1 & 0 \\ 0 & (0.01\text{ m/cm})^2\end{bmatrix}\begin{bmatrix}3\text{ m} - 2\text{ m} \\ 300\text{ cm} - 200\text{ cm}\end{bmatrix}$$

$$= [1\text{ m} \ \ 100\text{ cm}]\begin{bmatrix}1\text{ m} \\ 0.01\text{ m}^2/\text{cm}\end{bmatrix} = 1\text{ m}^2 + 1\text{ m}^2 = 2\text{ m}^2, \tag{3.4}$$

from which it follows that $\Delta d = \sqrt{2}$ m.

The above exercise with a trivial metric tensor is, obviously, a huge overkill for this simple artificial situation. As many complicated things, it becomes more useful as the complexity increases.

3.1.2   POLAR COORDINATES

One step up in complexity is obtained by the introduction of *curvilinear coordinates*. The basic example of such coordinates are *polar coordinates*.

In these coordinates, to mark the position of a point on a plain, rather than using two distances along two axes, we use one distance to the origin r, and one angle ϕ from some chosen zero direction. Most obviously the distance between arbitrary two points $(\rho_1; \phi_1)$ and $(\rho_2; \phi_2)$ may be found using the cosine theorem:

$$s^2 = \rho_1^2 + \rho_2^2 - 2\rho_1\rho_2 \cos(\phi_1 - \phi_2).$$

At first this appears to look nothing like the generalized Pythagorean theorem

$$\Delta s^2 = g_{11}\Delta\rho^2 + g_{11}\Delta\phi^2 + g_{12}\Delta\rho\Delta\phi + g_{21}\Delta\phi\Delta\rho.$$

However it is, for small triangles. Let us write the law of cosines not for two arbitrary points $(\rho_1; \phi_1)$ and $(\rho_2; \phi_2)$, but for two points that are very close to each other: $(\rho; \phi)$ and $(\rho + d\rho; \phi + d\phi)$:

$$ds^2 = \rho^2 + (\rho + d\rho)^2 - 2r(\rho + d\rho)\cos(\phi + d\phi - \phi) = d\rho^2 + r^2d\phi^2 + \rho d\rho d\phi^2$$

(the intermediate steps should be filled in by the reader). We now perform the usual procedure in calculus: reject all terms which have higher powers of very small differences (we have only one such term: $rdrd\phi^2$). Then the generalized Pythagorean theorem emerges:

$$ds^2 = d\rho^2 + \rho^2 d\phi^2. \tag{3.5}$$

We see that the metric tensor for the flat plane in polar coordinates is

$$\begin{bmatrix} 1 & 0 \\ 0 & \rho^2 \end{bmatrix}. \tag{3.6}$$

3.1.3   CURVED SURFACE

Now, let us look at a curved surface. Anyone who has ever tried to pack up an uninflated beach ball should be aware of the fact that it is not possible to flatten the surface of a sphere. The surface of a sphere has curvature, which makes it fundamentally different from a flat plane. It is not, really, possible to talk about straight lines on the surface of a sphere: observing the situation from the vantage point of the third dimension we realize that every line on the spherical surface is curved.

There are, however, special lines on the surface of any curved surface: the *great circles*. Empirically, we may obtain a section of a great circle between any two points on the surface of a sphere by stretching a string on between those two points (and making sure that the friction between the surface of the sphere and the string is negligible). In three dimensions, we may obtain great circles by intersecting the sphere by a plane that passes through the two points on the surface and the center of the sphere. If we consider Earth's geography, great circles correspond to the meridians (but not to the parallels: those are the *small circles*). They are obtained by intersecting the sphere by a plane which does *not* pass through the center.

A two-dimensional being confined to a surface of a sphere and unaware of the existence of the third dimension would experience great circles as straight lines. As a matter of fact, we experience great circles as straight lines when we travel on the Earth's surface.

So, to obtain the equivalent of Pythagorean theorem for the surface of sphere we may want to be considering the distances measured along the great circles. What does the generalized Pythagorean theorem look like in this case?

Before the 1960s, high school students were required to study spherical geometry and spherical trigonometry as a part of their regular mathematics courses. These are interesting subjects, with many applications in astronomy, navigation, and planetary science. The results look strange to those of us who are only familiar with flat plane geometry. For example, the sum of the angles of triangle made up by the sections of great circles is always greater than π radians (in other words, greater than 180°). The area of a triangle is always larger than the area of a corresponding flat triangle, and so on.

One of the theorems of spherical geometry is the "Spherical Pythagorean theorem." It states that for a right triangle drawn on the surface of a sphere of radius R, the following relation holds:

$$\cos\frac{a}{R}\cos\frac{b}{R} = \cos\frac{c}{R},$$

where, as usual, a and b are the lengths of the two sides adjacent the right angle, and c is the length of the hypotenuse. The sides of the spherical triangle in this theorem are, of course, the sections of great circles. At first sight, this theorem again does not look like the basic Pythagorean theorem. However, for very small right triangles, cosines can be approximated: $\cos a/R \approx = 1 - a^2/2R^2$, and the usual Pythagorean theorem results.

This result is an example of a very general statement, which is central both to the theory of curved spaces in mathematics, and to the theory of curved spacetimes in GR. For curved surfaces it says that locally (that is very close to the observer) any curved surface is, approximately, Euclidean.

However, the surface of a sphere is not Euclidean, it is curved. Is there a version of Pythagorean theorem that reflects this fact?

Yes, but it is more convenient to use the equivalent of polar coordinates to show this. A point on the surface of the sphere is chosen to be the origin, and one of great circles passing through the origin is chosen to be the direction 0. The coordinates of any other point are the distance ρ from the origin (along some great circle), and the angle ϕ between the zero direction great circle and the great circle that connects the origin with the point. This is almost identical to the longitude-latitude coordinates we give to locations on the Earth's surface. The only difference is that instead of measuring latitude from the equator in units of degrees, these coordinates measure co-latitude from the origin (the "pole") in units of length.

The metric tensor for the surface of a sphere in these units is

$$\begin{bmatrix} 1 & 0 \\ 0 & \sin^2(\frac{\rho}{R}) \end{bmatrix}, \tag{3.7}$$

where R is the radius of the sphere in three dimensions. For the small values of ρ the sin can be approximated by the value of its argument, and the flat plane metric in polar coordinates is again recovered.

3.1.4   GRAVITATIONAL WAVE

Let us, finally, express mathematically what happens in a propagating GW. As was explained at the end of the previous chapter, a GW distorts the space in the two orthogonal directions. The following metric tensor describes space which is both "stretched" in the x-direction, and "contracted" in the y-direction by the same small factor ϵ:

$$\begin{bmatrix} 1 + \epsilon & 0 \\ 0 & 1 - \epsilon \end{bmatrix}. \tag{3.8}$$

This is one of the two possible *polarizations* of such distortion. The other polarization is produced by stretching and contracting the plane along the "diagonal" directions, at 45° angles to the x- and y-axes. The metric tensor for this polarization is

$$\begin{bmatrix} 1 & \epsilon \\ -\epsilon & 1 \end{bmatrix}. \tag{3.9}$$

PART II

Electromagnetism and Gravitation

CHAPTER 4

Overview

At the level of introductory physics the laws of gravitation appear to be very similar to the laws of electrostatics: Coulomb's law

$$\mathbf{F} = \frac{1}{4\pi\epsilon_0}\frac{qQ}{r^2}\hat{\mathbf{r}}$$
(4.1)

and Newton's law of gravity

$$\mathbf{F} = -G\frac{mM}{r^2}\hat{\mathbf{r}}$$
(4.2)

have identical dependence on the distance, and so may both be written as

$$\mathbf{F} = k\frac{sS}{r^2}\hat{\mathbf{r}},$$
(4.3)

where the "sources" s and S could stand either for mass or for electrical charge, and constant "k" is to be either to $\frac{1}{4\pi\epsilon_0}$ or $k = -G = -6.673 \times 10^{-11}\frac{m^3}{kgs^2}$ depending on the source. If anything, gravity appears easier, since the interaction is always attractive, so there is no sign to worry about. Moreover, the fact that the mass m_g of Newton's gravity is identical with the inertial mass m_i makes matters even easier: both masses cancel out in innumerable simple problems leaving simpler looking answers.

The reality, of course, is that gravity is several orders more complicated than electrodynamics. This may be suspected already at the introductory level if one considers what the "interaction is always attractive" entails. From Newton's law it immediately follows that gravitational potential energy is given by

$$U_g = -G\frac{Mm}{r}.$$
(4.4)

Now suppose we bring two arbitrary masses M and m closer and closer together. Then, in principle, we can extract an infinite amount of energy. If we allow the two masses merge, we have a stronger source, and the possibility of extracting yet more energy.

The relativistic theory of gravity, GR, was designed almost from the outset to be a nonlinear theory, to ensure that the energy of the gravitational field also acts as its own source. The nonlinear terms in GR limit the amount of the total energy that can be extracted to a value of the order Σmc^2, the rest mass energy of the masses. This "not bug but a feature" of GR makes it an extremely difficult theory to learn even if it were the only complication compared to electrodynamics.

There are other complications, however. Electrodynamics can be described with two vector fields: the electric field $\mathbf{E}$ and the magnetic field $\mathbf{B}$. These two fields are produced by the sources: electric charge density ρ and electric current $\mathbf{J}$. On a more sophisticated level, using special relativity, electric charge and current may be combined into a single *current density 4-vector* $J^\mu = (c\rho, \mathbf{J})$. This 4-vector gives rise to the combination of spatial and time derivatives of *4-vector potential* $A^\mu = (V/c, \mathbf{A})$, and particular combinations of these derivatives are then identified with the electric and magnetic fields:

All of the analogous quantities in theory of gravity obtain more components and become higher order tensors. The relativistic mass-energy density is the time-time component of the second-rank *stress–energy tensor* $T^{\mu\nu}$. The number of potentials increases to ten (the components of the *metric tensor* $g^{\mu\nu}$, and to fully represent the fields one has to work with the fourth-rank *curvature tensor* $R^{\alpha\beta\gamma\delta}$.

Thus, GR contains massive complications compared to the classical electrodynamics. It is a non-linear theory of second-rank tensor fields. The number of closed-form solutions for this theory is extremely limited (on the order of one), and not many more numerical solutions have existed until very recently. The field of numerical relativity has experienced an explosive growth with the growth of computer technology, and has brought with it new understanding and new visualization.

It is not the purpose of this book to teach GR. Instead of focusing on the complexity of the theory, it aims to discuss the similarity of gravity to the classical EM theory in a limited number of situations, before going on to an overview of the experimental attempts of the last few decades.

To this end, we will be dealing with *linearized* gravity. In this simplification of the full GR gravitational fields are small enough that we may neglect the fact that their own energy acts as an additional source. Even though we still have to grapple with the higher-order tensors, the complexity of the theory is greatly reduced, and the theory may be made very similar to the Maxwell's electrodynamics.

CHAPTER 5

Static Sources and Static Multipoles

5.1 STATIC SOURCES

We begin the a discussion of a couple of three-dimensional tensors widely useful in classical mechanics and electrodynamics. As will become obvious later, these tensors have immediate applicability to the subject of gravitation waves.

We explore the fields and potentials produced by a localized distribution of sources. Since we are dealing with weak field approximation in the case of gravity, we may use Eq. (4.3). Since this force is conservative, we may introduce potential ϕ. In terms of this potential, both Newton's law of gravitation and Coulomb law of electrostatics can be expressed through *Poisson equation*:

$$\nabla^2 \phi = -4\pi k \rho, \tag{5.1}$$

where ϕ is is generated by the source density ρ. As before, the constant "k" is to be either to $\frac{1}{4\pi\epsilon_0}$ or $k = -G$ depending on the source.

The solution of this equation at any moment in time t and at any location $\mathbf{r}_d$ in 3-d space may be immediately written in a general form as the linear supersposition of $1/r$ potentials produced by each element of the source:

$$\phi(\mathbf{r}_d) = k \int \frac{\rho(\mathbf{r}_s)}{|\mathbf{r}_s - \mathbf{r}_d|} d^3 r_s. \tag{5.2}$$

The subscripts "s" and "d" refer to the "source" and to the "detector." It should be kept in mind that potentials are not directly observable in either the electromagnetic or gravitational case. To obtain the observable fields, gradients of the potentials will have eventually to be taken.

We will often be interested in the potentials and fields produced by "point" sources (e.g., stars located mega-parsecs away). The most convenient way of dealing with such point sources is through the use of a three-dimensional δ-function. So, to obtain the point source of magnitude S located at a point $\mathbf{r}_s$, we write the source density as $\rho(\mathbf{r}) = S\delta^3(\mathbf{r} - \mathbf{r}_s)$. Once integrated, the expression with such source density produce correct results for point sources.

The integration in general is performed over the entirety of space, that is over all the sources in the universe. We are interested only in the potential produced by a few sources within some bounded region (this naturally excludes such useful physical idealizations as infinite or half-infinite planes or wires).

The potential produced by such a localized distribution may be conveniently expanded into a series, whose terms fall off with the distance as consecutive inverse powers of distance between the detector and some single point within the source:

$$\phi(\mathbf{r}_d) = \sum \frac{\phi_n(\hat{\mathbf{r}}_d)}{r^n}.$$

The ϕ_n terms in this series are *not* spherically symmetric. This is represented by showing $\phi_n = \phi_n(\hat{\mathbf{r}}_d)$: each of the ϕ_n is given by a formula, which does not contain the distance, but contains the angle from the origin to the detector. This is to be expected: the potential produced by a source is not spherically symmetric unless the source itself is spherically symmetric. Since we aim to describe the potential produced by several point sources, themselves not necessarily neatly spherically arranged, we should expect to see the angular dependence in the terms ϕ_n of the above series. Thus the dependence of the potential on the distance from the origin has been separated into the $1/r^n$ terms, however the dependence on the direction is contained in each term.

As n increases, the angular dependence $\phi_n(\hat{\mathbf{r}}_d)$ becomes progressively more complicated. The angular pattern $\hat{\mathbf{r}}_d$ depends on the source through the *multipole moments M_n*.

5.2   MULTIPOLE EXPANSION

Mathematically, the angular dependence in this and later sections comes from the fact that the distance from the origin to the detector r_d does not equal the distance from an element of the source to the detector. In this and the next chapters it is useful to keep in mind the following picture:

As the drawing shows, the "source" is a "binary" composed of two objects (red and blue), which are meant to represent two astronomical objects, such as black holes or neutron stars. The total size of the source is the average separation of the two objects. The typical size varies from a few million down to zero kilometers. The origin of the coordinate system is located at the center of mass of that binary. The "detector" is on Earth which is many megaparsecs away from the origin.

The distance $|\mathbf{r}_d - \mathbf{r}_s|$ can be expressed exactly through the theorem of cosines:

$$|\mathbf{r}_d - \mathbf{r}_s| = \sqrt{r_d^2 + r_s^2 - 2\mathbf{r}_d \cdot \mathbf{r}_s}, \tag{5.3}$$

and since r_d is many orders of magnitude larger than r_s it is sensible to expand this distance in terms of small quantity r_s/r_d:

$$|\mathbf{r}_d - \mathbf{r}_s| = r_d \sqrt{1 + \left(\frac{r_s}{r_d}\right)^2 - \frac{2\hat{\mathbf{r}}_d \cdot \mathbf{r}_s}{r_d}}. \tag{5.4}$$

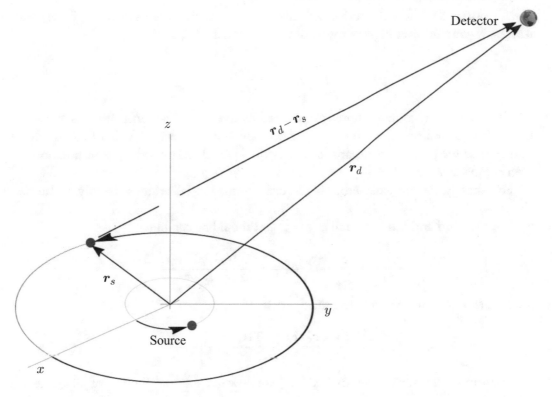

Figure 5.1: The source and the detector.

To understand this angular dependence we expand the inverse distance $1/|\mathbf{r}_s - \mathbf{r}_d|$, either by applying the binomial theorem, or Taylor expansion. This leads to

$$|\mathbf{r}_d - \mathbf{r}_s|^{-1} = (r_d^2 + r_s^2 - 2\mathbf{r}_d \cdot \mathbf{r}_s)^{-1/2} =$$

$$r_d^{-1} \left[1 - \frac{1}{2} \left(\frac{r_s^2}{r_d^2} - \frac{2\hat{\mathbf{r}}_d \cdot \mathbf{r}_s}{r_d} \right) + \frac{3}{8} \left(\frac{r_s^2}{r_d^2} - \frac{2\hat{\mathbf{r}}_d \cdot \mathbf{r}_s}{r_d} \right)^2 + \dots \right] = \tag{5.5}$$

$$\frac{1}{r_d} + \frac{\mathbf{r}_s \cdot \hat{\mathbf{r}}_d}{r_d^2} - \frac{r_s^2}{2r_d^3} + \frac{3}{8r_d} \frac{4(\mathbf{r}_s \cdot \hat{\mathbf{r}}_d)^2}{r_d^2} + \dots$$

We may already see some familiar terms: the first one, $1/r_d$, falls proportionally to the distance from the origin (which itself is located somewhere within the source), the second one is inversely proportional to $1/r_d^2$ and contains some angular dependence through the dot product $\mathbf{r}_s \cdot \hat{\mathbf{r}}_d$.

We notice also how the series will involve progressively more terms: as the expansion continues, it contains ever higher powers of the expression

$$\left(\frac{r_s^2}{r_d^2} - \frac{2\hat{\mathbf{r}}_d \cdot \mathbf{r}_s}{r_d} \right).$$

Each next term contains more complicated features. The first term, $1/r_d$, is just a scalar quantity. The second one is written as a scalar product of two vectors. What follows after the second term is the part most relevant for this book. It requires 2nd order tensor notation to be written compactly.

Fortunately, the more numerous highe-order terms also fall off more rapidly, and are wery rarely useful.

Both the 3rd and 4th terms fall as $1/r_d^3$, and should be combined

$$-\frac{r_s^2}{2r_d^3} + \frac{3}{8r_d}\frac{4(\mathbf{r}_s \cdot \hat{\mathbf{r}}_d)^2}{r_d^2} = \frac{3}{2}\frac{(\mathbf{r}_s \cdot \hat{\mathbf{r}}_d)^2 - r_s^2/3}{r_d^3}$$

to produce an equation correct to the third order:

$$r^{-1} = \frac{1}{r_d} + \frac{\mathbf{r}_s \cdot \hat{\mathbf{r}}_d}{r_d^2} + \frac{3}{2}\frac{(\mathbf{r}_s \cdot \hat{\mathbf{r}}_d)^2 - r_s^2/3}{r_d^3}. \qquad (5.6)$$

Performing the summation in Eq. (5.2) we now get the contribution of all elements of the source:

$$\phi(\mathbf{r}_d) = -k \int \frac{\rho(\mathbf{r}_s)}{r} = -k \int \rho(\mathbf{r}_s)\left[\frac{1}{r_d} + \frac{\mathbf{r}_s \cdot \hat{\mathbf{r}}_d}{r_d^2} + \frac{3}{2}\frac{(\mathbf{r}_s \cdot \hat{\mathbf{r}}_d)^2 - r_s^2/3}{r_d^3} \right] d^3 r_s. \qquad (5.7)$$

We have obtained the first three *Cartesian multipole moments*.

The leading *monopole* term in the expansion has $n = 1$ and corresponds to the basic static potential produced by the net source:

$$M_1 = \int \rho(\mathbf{r}_s)d^3 r_s.$$

This is the only spherically symmetric term in the expansion. It does not "care" about the geometry of the source. When the source itself is perfectly spherically symmetric, the entire potential produced by the source is describable by the monopole term. The potential due to this term alone is identical to the potential that would be produced by a point charge located at the center of the spherical charge distribution.

(Note: it is necessary that the origin is placed at the center of the spherical source for this description to work. If, for some reason, the origin is chosen to be located elsewhere, all other terms in the expansion will have to be used to produce a spherically symmetric field which is *not* centered at the origin).

The 2nd term corresponds to the *dipole* potential

$$M_2(\mathbf{r}_d) \equiv \mathbf{p} = \int \rho(\mathbf{r}_s)\mathbf{r}_s d^3 r_s.$$

This is the dipole moment. It is a vector, which measures how far and in which direction the positive and negative charges within the source are "separated." The "standard" example of the dipole source is a pair of electrical charges of the opposite signs $\pm q$ located symmetrically at $z = \pm d/2$, where the distance d between the charges is decreased to zero at the same time as the magnitudes of charges q are increased so as to keep the magnitude of the *dipole moment* constant.

In the situations where only one sign of charge is present the dipole moment may always be made equal to zero. For instance, a pair of unequal charges q_1 and $q_2 \neq q_1$ of the same sign located symmetrically at $z = \pm d/2$ may be expressed as a combination of a monopole and a dipole. However, if we reposition the origin to the *barycenter* of the charges, so that q_1 is located at $z_1 = a \frac{q_2}{q_1+q_2}$ and q_2 is at $z_2 = a \frac{q_1}{q_1+q_2}$, the dipole moment of the system with respect to the new origin is zero. We will calculate this and other results when discuss the "static binary" in a few paragraphs.

Mathematically, the dipole moment is obtained by performing a vector sum of a number of vectors of different magnitudes.

The 3rd term in the expansion corresponds to the *quadrupole* moment.

$$M_3 \equiv Q = \int S_i [3r_{si}r_{sj} - r_{si}^2 \delta_{ij}] d^3 r_s. \tag{5.8}$$

This is not a single number nor a vector, but a combination of 9 numbers, corresponding to different values of i and j. In other words, it is a rank-2 tensor representing the 2nd-order moment of source distribution. Not all 9 numbers are different: there is a symmetry between the i and j, so there are only 6 different numbers. Just as the dipole moment vector has the magnitude which is proportional to the "separation" between charges, the quadrupole moment also has the magnitude, which measures whether an object spherical, *prolate* (elongated compared to a sphere), or *oblate* (flattened) (Fig. 5.2).

The quadrupole is "built around" integral of $\rho(\mathbf{r}_s)r_s^2$. One may be familiar with a different quantity built around the same integral: an object's moment of inertia. It is worthwhile to compare and contrast the two:

$$Q_{ij} \equiv \int [3r_{si}r_{sj} - r_s^2\delta_{ij}]\rho(\mathbf{r}_s)d^3 r \tag{5.9}$$

$$I_{ij} \equiv -\int [r_{si}r_{sj} - r_s^2\delta_{ij}]\rho(\mathbf{r}_s)d^3 r. \tag{5.10}$$

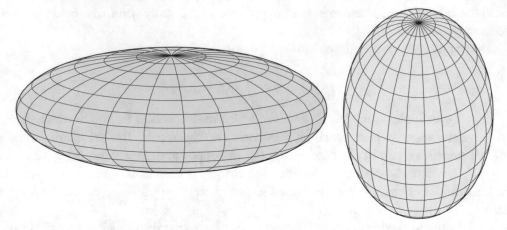

Figure 5.2: Oblate and prolate spheroids.

In the case of moment of inertia the source density ρ stands solely for the inertial mass density. The examination of the two expressions shows that there is a strong similarity between them. In fact, the moment of inertia tensor can be used to express the gravitational potential instead of the quadrupole tensor. The *MacCullagh's formula* for the gravitational potential [5]

$$V = -\frac{GMm}{r} + \frac{GM}{2r^3}[3I_r - (I_1 + I_2 + I_3)]$$

is widely used in the planetary science.

It is more convenient, though, to use the inertia tensor to describe the rotational proper-
ties of an object, and quadrupole tensor to describe the "field-creating" properties. Because the two tensors are used to describe different properties of an object, they have different properties themselves. Let us look at a few simple examples.

5.2.1 SPHERE

A solid sphere of radius R and uniform mass m has a surface described by

$$x^2 + y^2 + z^2 = R^2.$$

The moment of inertia tensor and quadrupole moment tensor are

$$I_{sphere} = \frac{2mR^2}{5}\begin{bmatrix} 1 & 0 & 0 \\ 0 & 1 & 0 \\ 0 & 0 & 1 \end{bmatrix} \qquad\qquad Q_{sphere} = \mathbf{0}.$$

As was mentioned above, a common adage is that quadrupole moment tells us how aspherical an object is. A sphere is not aspherical at all, so naturally its quadrupole moment is zero.

However, it obviously has rotational inertia when rotated around any of its diameters, hence it has non-zero symmetric tensor of rotational inertia.

5.2.2 CYLINDER

A solid cylinder of radius R and height h has the moment of inertia

$$I_{cylinder} = \frac{m}{12} \begin{bmatrix} 3R^2 + h^2 & 0 & 0 \\ 0 & 3R^2 + h^2 & 0 \\ 0 & 0 & 6R^2 \end{bmatrix},$$

and the quadrupole moment tensor is:

$$Q_{cylinder} = \frac{m}{12} \begin{bmatrix} h^2 - 3R^2 & 0 & 0 \\ 0 & h^2 - 3R^2 & 0 \\ 0 & 0 & 6R^2 - 2h^2 \end{bmatrix}.$$

The diagonal entries of the moment of inertia tensor are all positive. In fact, that is always true: any 3-dimensional object has rotational inertia around any axis.

The quadrupole moment tensor has both positive and negative entries. In fact, the sum of the diagonal entries of any quadrupole tensor is always *zero*. The *trace* of a tensor is the sum of its diagonal entries. Any quadrupole tensor is always *traceless*.

If a matrix has positive diagonal entries, it can be used to describe an 3-dimensional geometrical ellipsoid. This *ellipsoid of inertia* has the smoothed out outlines of the original shape.

It is not possible to describe a traceless tensor using an ellipsoid. Rather, quadrupole tensor describes how much would have to be "added to" or "taken away from" a uniform sphere along each axis to obtain the smoothed out version of the original object.

5.2.3 CUBE

(See [4])

$$I_{cube} = \frac{ml}{6} \begin{bmatrix} 1 & 0 & 0 \\ 0 & 1 & 0 \\ 0 & 0 & 1 \end{bmatrix} \qquad\qquad Q_{cube} = \mathbf{0}.$$

Here we see the limits of the "common adage" (quadrupole moment tells us how aspherical an object is). A cube is obviously not a sphere. However it is neither prolate nor oblate, and so the quadrupole moment of a cube can tell us nothing about its asphericity. To show how it is different from a sphere, we would need to calculate hight multipole moments.

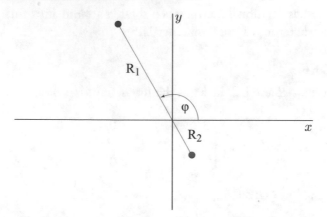

Figure 5.3: A stationary binary.

5.2.4   FIRST MULTIPOLE MOMENTS OF A BINARY

Let us calculate several multipole moments of a (Fig. 5.3) "binary": just two point sources placed at some distance $L = r_1 + r_2$ from each other.

Specifically, let us place the origin of the coordinate system somewhere between the positions of the two sources (not necessarily in the middle). Then the coordinates of the two sources are

$$
\begin{bmatrix} x_1 \\ y_1 \\ z_1 \end{bmatrix} = \begin{bmatrix} r_1 \cos\phi \\ r_1 \sin\phi \\ 0 \end{bmatrix}
\tag{5.11}
$$

and

$$
\begin{bmatrix} x_2 \\ y_2 \\ z_2 \end{bmatrix} = \begin{bmatrix} r_2 \cos(\pi + \phi) \\ r_2 \sin(\pi + \phi) \\ 0 \end{bmatrix}.
\tag{5.12}
$$

Dipole Moment

$$
\begin{aligned}
\begin{bmatrix} p_x \\ p_y \\ p_z \end{bmatrix} &= \begin{bmatrix} S_1 r_1 \cos\phi \\ S_1 r_1 \sin\phi \\ 0 \end{bmatrix} + \begin{bmatrix} S_2 r_2 \cos(\pi + \phi) \\ S_2 r_2 \sin(\pi + \phi) \\ 0 \end{bmatrix} \\
&= \begin{bmatrix} S_1 r_1 \cos\phi + S_2 r_2(-\cos\phi) \\ S_1 r_1 \sin\phi + S_2 r_2(-\sin\phi) \\ 0 \end{bmatrix} = (S_1 r_1 - S_2 r_2) \begin{bmatrix} \cos\phi \\ \sin\phi \\ 0 \end{bmatrix}.
\end{aligned}
\tag{5.13}
$$

This is exactly what we should expect for the dipole moment: it is a vector, whose direction is the same as the direction of the line connecting the two point sources.

If the origin were placed in the middle so $r_1 = r_2 = L/2$ then the magnitude of the dipole moment is $(S_1 - S_2)L/2$. If we set $S_1 = +q$ and $S_2 = -q$ then we get a simple dipole: $p = qL$. If, on the other hand, the two sources are of the same sign, then the magnitude is zero.

Perhaps, surprisingly, if we take two source of the same magnitude and sign, but set the origin not in the middle between them then the dipole moment is calculated to be not zero. However, this is as it should be. The multipole moment coefficients depend on the choice of origin. Even a single point source, if placed not at the origin of coordinate system, possesses dipole moment (as well as quadrupole, and, in fact, all higher multipole moments).

5.2.5   CENTER OF MASS

It is useful for later to note that the dipole moment becomes zero if the origin is set so that $S_1 r_1 = S_2 r_2$. If we are talking about the gravitational source, with $S_1 = m_1$ and $S_2 = m_2$, then $m_1 r_1 = m_2 r_2$ defines the location of the center of mass of the system. Using $r_1 + r_2 = L$ and defining the total mass of the binary $m = m_1 + m_2$ we can use simple algebra to find

$$ r_1 = \frac{m_2}{m} L \qquad\qquad r_2 = \frac{m_1}{m} L. \qquad (5.14) $$

The center of mass is an important concept in GWs. One of the postulates of GR is the equivalence of gravitational and inertial masses. Since any self-contained system moves around its own center of mass, the gravitational dipole moment of any self-contained system is zero.

Second Moment

$$
\begin{bmatrix} I_{xx} & I_{xy} & I_{xz} \\ I_{yx} & I_{yy} & I_{yz} \\ I_{zx} & I_{zy} & I_{zz} \end{bmatrix}
$$

$$
= \begin{bmatrix} S_1(r_1\cos\phi)(r_1\cos\phi) & S_1(r_1\cos\phi)(r_1\sin\phi) & 0 \\ S_1(r_1\sin\phi)(r_1\cos\phi) & S_1(r_1\sin\phi)(r_1\sin\phi) & 0 \\ 0 & 0 & 0 \end{bmatrix} +
$$

$$
\begin{bmatrix} S_2(r_2\cos(\pi+\phi))(r_2\cos(\pi+\phi)) & S_2(r_1\cos(\pi+\phi))(r_2\sin(\pi+\phi)) & 0 \\ S_2(r_2\sin(\pi+\phi))(r_2\cos(\pi+\phi)) & S_2(r_2\sin(\pi+\phi))(r_2\sin(\pi+\phi)) & 0 \\ 0 & 0 & 0 \end{bmatrix}
$$

$$
= (S_1 r_1^2 + S_2 r_2^2) \begin{bmatrix} \cos^2\phi & \sin\phi\cos\phi & 0 \\ \sin\phi\cos\phi & \sin^2\phi & 0 \\ 0 & 0 & 0 \end{bmatrix}. \qquad (5.15)
$$

We have obtained a 2nd rank tensor. The magnitude is what we would calculate as the moment of inertia of two points in elementary physics. This quantity is positive if both sources are positive. Anything containing the z-coordinate is zero, since we chose the coordinate system so that the sources are in the xy-plane.

If the origin were placed in the middle so $r_1 = r_2 = L/2$ then the magnitude of the second moment is $(S_1 + S_2)L^2/4$. In case of equal sources, $S_1 = S_2 = S$, we obtain the 2nd moment of a dumbbell, $I = 2S(L/2)^2$.

5.2.6   REDUCED MASS

Let us specify that we are considering gravitational interaction, and set the masses of the objects as the sources: $S_1 = m_1$ and $S_2 = m_2$. Let us set the origin again at the center of mass of the binary. The magnitude becomes

$$I = m_1 r_1^2 + m_2 r_2^2 = \frac{m_1 m_2^2 L^2 + m_2 m_1^2 L^2}{M^2} = \frac{2m_1 m_2}{M} L^2. \tag{5.16}$$

The expression $m_1 m_2/M$ is ubiquitous in the areas of physics that deals with two-body problems. This expression defines the *reduced mass*

$$\mu \equiv \frac{m_1 m_2}{M}. \tag{5.17}$$

In terms of the reduced mass then the 2nd moment of a binary is just

$$I = \mu L^2 \begin{bmatrix} \cos^2 \phi & \sin \phi \cos \phi & 0 \\ \sin \phi \cos \phi & \sin^2 \phi & 0 \\ 0 & 0 & 0 \end{bmatrix}. \tag{5.18}$$

Quadrupole Moment and Moment of Inertia around the Center of Mass

From the definitions in Eqs. (5.9) and (5.10) we can now calculate the quadrupole moment, which will be useful to us later:

$$Q = \mu L^2 \begin{bmatrix} \cos^2 \phi - \frac{1}{3} & \sin \phi \cos \phi & 0 \\ \sin \phi \cos \phi & \sin^2 \phi - \frac{1}{3} & 0 \\ 0 & 0 & -\frac{1}{3} \end{bmatrix}, \tag{5.19}$$

and the moment of inertia (which we will not need):

$$I = \mu L^2 \begin{bmatrix} \sin^2 \phi & -\sin \phi \cos \phi & 0 \\ -\sin \phi \cos \phi & \cos^2 \phi & 0 \\ 0 & 0 & 1 \end{bmatrix}. \tag{5.20}$$

We can again easily see meaning of these two tensors if we set ϕ to zero (so both sources are positioned on the x-axis, then

$$Q = \mu L^2 \begin{bmatrix} 2 & 0 & 0 \\ 0 & -1 & 0 \\ 0 & 0 & -1 \end{bmatrix} \quad \text{and} \quad I = \mu L^2 \begin{bmatrix} 0 & 0 & 0 \\ 0 & 1 & 0 \\ 0 & 0 & 1 \end{bmatrix}.$$

The Q-tensor shows that the object is elongated (prolate) along the x-axis, and compressed (oblate) along both y- and z-axes, while the I-tensor shows rotational inertia of μr_s^2 when rotated around either the y- or the z-axes. The moment of inertia for the rotation around the x-axis in this case is (obviously) zero.

CHAPTER 6

Waves from Retarded Potentials

6.1   OVERVIEW

In the previous chapter we have obtained a series expansion of the static potentials into a series of terms falling as increasing powers of $1/r$. To obtain *waves* we have to expand the discussion to *time-dependent* behaviors. In this chapter we will do that in the simplest possible way: by introducing time delay to the propagation of the potential.

The potential with the built in time delay automatically satisfies the wave equation. However we will not be satisfied with simply stating that fact. The measurable quantities in electromagnetism are not potentials, but gradients of potentials: fields. To discover that a charge is radiating, it is necessary to calculate these fields.

The situation is even more complex in case of gravity. In the GR the measurable quantities are obtained through a complicated combination of *gradients of gradients* of the potentials. Before we discuss that full theory in a later chapter we show the results in the simple settings. In this chapter we show the expected effect of the time delay on the first gradient field.

6.2   TIME DELAY

If the Poisson equation

$$\nabla^2 \phi = -4\pi k \rho, \tag{6.1}$$

where the fully correct description of any physical phenomenon, any change in the source at time t would result in the change in the potential at the location of the detector at the same moment in time t. That is, the static equation, e.g., Newton's law of gravitation, implies instantaneous propagation of signals. This may be corrected by postulating different times t_d and t_s for the times of "detection" and the times of "change at the source," respectively.

$$\phi(\mathbf{r}_d, t_d) = k \int \frac{\rho(\mathbf{r}_s, t_s)}{|\mathbf{r}_s - \mathbf{r}_d|} d^3 r_s. \tag{6.2}$$

Assuming the changes propagate at a finite velocity c, it takes $|\mathbf{r}_d - \mathbf{r}_s|/c$ units of time for the changes to propagate from the source to the detector. Therefore,

$$t_d = t_s + |\mathbf{r_d} - \mathbf{r_s}|/c, \tag{6.3}$$

and the potential at the detector now has to be calculated by integrating over *retarded* potentials:

$$\phi(\mathbf{r}_d, t_d) = k \int \frac{\rho(\mathbf{r}_s, t_d - |\mathbf{r}_s - \mathbf{r}_d|/c)}{|\mathbf{r}_s - \mathbf{r}_d|} d^3 r_s \qquad (6.4)$$

as may be checked by direct calculation [2]. With this modified equation, the potential ϕ satisfies the wave equation.

6.3 FIELD AS THE GRADIENT OF POTENTIAL

6.3.1 VARIATIONS IN TIME AND VARIATIONS IN SPACE

In the previous chapter we saw that by expanding the $|\mathbf{r}_s - \mathbf{r}_d|$ in the denominator the potential can be written as the sum of the successive powers of inverse distance. The first term in that expansion falls off as $1/r$. The gradient of that term, which corresponds to the field, falls off as $1/r^2$, which corresponds to the Newton's law of gravity (or Coulomb's law of electrostatics).

The radiation *field itself* falls off as $1/r$. This result does not come from expanding the $|\mathbf{r}_s - \mathbf{r}_d|$ in the denominator: nothing can be obtained from that expansion that we did not obtain in the previous chapter. In fact, we just work with the first term of the expansion of the previous chapter. That simplifies Eq. (6.4) to

$$\begin{aligned} \phi(\mathbf{r}_d, t_d) &= k \int \frac{\rho(\mathbf{r}_s, t_d - |\mathbf{r}_s - \mathbf{r}_d|/c)}{r_d} d^3 r_s \\ &= \frac{k}{r_d} \int \rho(\mathbf{r}_s, t_d - |\mathbf{r}_s - \mathbf{r}_d|/c) d^3 r_s. \end{aligned} \qquad (6.5)$$

We now find the gradient field:

$$-\nabla \phi(\mathbf{r}_d, t_d) = -\frac{k}{r_d} \nabla \int \rho(\mathbf{r}_s, t_d - |\mathbf{r}_s - \mathbf{r}_d|/c) d^3 r_s. \qquad (6.6)$$

As the detector can see, the time of emission $t_s = t_d - |\mathbf{r}_s - \mathbf{r}_d|/c$ is a function of both time and position. The derivatives (spatial or time) do not keep the combination $t_d - |\mathbf{r}_s - \mathbf{r}_d|/c$ intact, "variations with time at the source are translated into variations in space" [3].

We use the chain rule for partial derivatives to obtain the gradient of the space source density as it is changing in time:

$$\nabla \rho = \frac{\partial \rho}{\partial t} \nabla (t_d - \frac{|\mathbf{r}_s - \mathbf{r}_d|}{c}) = -\frac{\nabla |\mathbf{r}_s - \mathbf{r}_d|}{c} \frac{\partial \rho}{\partial t} = -\hat{\mathbf{r}} \frac{1}{c} \frac{\partial \rho}{\partial t} \qquad (6.7)$$

and put it into Eq. (6.6) to obtain

$$-\nabla \phi = -\frac{k}{r_d} \int (\nabla \rho) d^3 r_s = \frac{k}{r_d} \int \frac{\hat{\mathbf{r}}}{c} \frac{\partial \rho}{\partial t} d^3 r_s = \hat{\mathbf{r}} \frac{k}{c r_d} \int \frac{\partial \rho}{\partial t} d^3 r_s. \qquad (6.8)$$

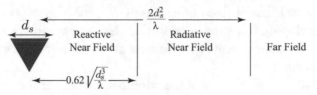

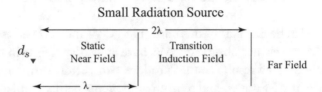

Figure 6.1: The common definitions of the three radiation zones for large and small sources.

We end up with the *field* which falls off as $1/r_d$. The expression for this field contains the combination $\dot{\rho}/c$. The dimensions of this combination are S/L, where S is the strength of the source (Coulombs or kilograms). The inverse of this, $c/\dot{\rho}$ may naturally be identified with the wavelength λ of produced radiation:

$$c/\dot{\rho} \equiv \lambda. \tag{6.9}$$

6.3.2   ZONES

The wavelength λ may be larger or smaller than the size of the source d_s. For artificially produced EM waves this leads to a distinction between "electromagnetically short" and "electromagnetically long" antennas. The situations of "long" antennas, $d_s > \lambda$, are more typical of human-engineered directional sources. Compact radiation sources, for which $d_s << \lambda$ are often fundamental physical objects (Fig. 6.1). In either case it is possible to consider different spatial regions or zones based on the distance $|\mathbf{r}_d - \mathbf{r}_s|$ between the source and the detector. A typical definition for compact sources is:

$$
\begin{aligned}
\text{Near (static) region:} \quad & |\mathbf{r}_d - \mathbf{r}_s| << \lambda \\
\text{Transition (induction) region:} \quad & \lambda < |\mathbf{r}_d - \mathbf{r}_s| < 2\lambda \\
\text{Far (radiation) region:} \quad & 2\lambda << |\mathbf{r}_d - \mathbf{r}_s|.
\end{aligned} \tag{6.10}
$$

In either near or transition regions we would have to include the effects of the static multipole terms calculated in the previous chapter. The approximation that led from Eq. (6.4) to Eq. (6.5) is valid only in the far region.

6.3.3   IMPOSSIBLE WAVES

We thus obtained a wave behavior from either Newtonian law of gravity or from Coulomb law. We will continue developing this expression in this chapter, and obtain many reliable results. However let us first note what is manifestly *incorrect* about the result above, as well as about all other results of this chapter.

Every expression for the gradient of the potential, that is, for the radiation field, has the direction $\hat{\mathbf{r}}$ (or opposite). This is the direction toward or away from the source of radiation. That is, all waves obtained in this way are the *longtitudonal* waves, similar to the compression sound waves.

As was discussed in he first part of this book, both GW and EM waves are *transverse* waves. Neither observed EM waves, nor observed GWs have any components in the direction of their propagation. Properly derived results should produce two vector components perpendicular to the $\hat{\mathbf{r}}$ direction in case of EM waves, and tensor components perpendicular to the $\hat{\mathbf{r}}$ direction in case of GWs.

However, the proper discussion of EM waves requires the consideration of magnetic fields in addition to the electric fields. The proper discussion of GW requires the consideration of many components of the full gravitaty-related tensors. We will develop these complications in later chapters.

6.4   MULTIPOLE RADIATION

Having obtained the waves from a single scalar potential, we examine the nature of waves that may be produced by a localized scalar source. We will expand the time-dependent potential, then apply Eq. (6.7) to obtain the fields.

In dealing with *static* multipoles, the elements of the source located at distances $r \equiv |\mathbf{r}_s - \mathbf{r}_d| \neq |r_d|$ (that is, close to the origin, but not *at* the origin) created multipole potentials, which fall with distance as $1/r_d^n$, where n is the multipole order. Since obtaining the fields from potentials requires taking derivatives with respect to distance, the fields associated with these potentials, fall with distance as $1/r_d^{n+1}$.

The situation is different for multipole radiation from localized sources. We are going to find that the fields fall with distance as $1/r_d$ for *every* multipole order. All terms which decrease faster than $1/r$ have been neglected. The tell-tale sign of radiation is the $1/r$ reduction in the field strength.

Each nth multipole term still has the general form of $\int \rho r_s^n$. Moreover, with the increasing multipole order the overall strength of the contribution decreases. However, the dependence on the distance remains of the form $1/r_d$.

To obtain multipole expansion of radiation we again expand $|\mathbf{r}_d - \mathbf{r}_s|$ around $|r_d|$. However this expansion is will now be done not in the denominator, but within the charge density ρ.

Instead of expanding the inverse distance $|\mathbf{r}_d - \mathbf{r}_s|^{-1}$ as we did for static multipoles, we expand the distance $|\mathbf{r}_d - \mathbf{r}_s|$ itself, neglecting even the r_s^2 part:

$$|\mathbf{r}_d - \mathbf{r}_s| = (r_d^2 + r_s^2 - 2\mathbf{r}_d \cdot \mathbf{r}_s)^{1/2} \approx r_d - \hat{\mathbf{r}}_d \cdot \mathbf{r}_s. \qquad (6.11)$$

The source density can be expressed in terms of the source density at the origin:

$$\rho(\mathbf{r}_s, t_d - |\mathbf{r}_d - \mathbf{r}_s|/c) = \rho(\mathbf{r}_s, t_d - r_d/c) - \frac{\hat{\mathbf{r}}_d \cdot \mathbf{r}_s}{c} \frac{\partial \rho(\mathbf{r}_s, t_d - r_d/c)}{\partial t}$$
$$+ \frac{1}{2!}(\frac{\hat{\mathbf{r}}_d \cdot \mathbf{r}_s}{c})^2 \frac{\partial^2 \rho(\mathbf{r}_s, t_d - r_d/c)}{\partial t^2} + \dots \qquad (6.12)$$

Putting this into the definition of the retarded potential Eq. (6.4), we obtain a succession of terms all of which fall inversely with the distance:

$$\phi_{monopole}(\hat{\mathbf{r}}_d, t_d) = \frac{k}{r_d} \int \rho(\mathbf{r}_s, t_d - r_d/c) d^3 r_s \qquad (6.13)$$

$$\phi_{dipole}(\hat{\mathbf{r}}_d, t_d) = \frac{k}{r_d} \int \frac{r_s}{c} \frac{\partial \rho(\mathbf{r}_s, t_d - r_d/c)}{\partial t} d^3 r_s \qquad (6.14)$$

$$\phi_{quadrupole}(\hat{\mathbf{r}}_d, t_d) = \frac{k}{r_d} \int \frac{1}{2!}(\frac{r_s}{c})^2 \frac{\partial^2 \rho(\mathbf{r}_s, t_d - r_d/c)}{\partial t^2} d^3 r_s \qquad (6.15)$$

and so on. If these were static potentials, taking the gradient to obtain fields would reduce the distance dependence to $1/r^2$. However, as we saw in the previous section, taking the gradient of the *retarded* potentials preserves the $1/r$ behavior for the fields.

The monopole term contains the integral of the total source. The potential produced by this term the familiar static potential. The application of ordinary spatial gradient to this potential produces the electrostatic field. But to obtain the time-varying wave fields we have to apply Eq. (6.7):

$$\nabla \phi \approx \hat{\mathbf{r}} \frac{1}{c} \frac{\partial \phi}{\partial t} = -\hat{\mathbf{r}} \frac{k}{rc} \frac{\partial}{\partial t} \int (\rho_0) d^3 r_s. \qquad (6.16)$$

This can be non-zero only if $\int (\rho_0) d^3 r_s$ changes with time. This is not possible either in case of electric charge, or mass. This means that there are no waves produced by the monopole sources.

The second term contains the first time derivative of the dipole moment. The time change in the dipole moment produces contributions to the potential.

The dipole moment of source distribution behaves differently in the case of electric charges than in the case of masses. The electric dipole moment, $\int (q r_s) d^3 r_s$ may perfectly well be variable in time, in fact that is the normal state of affairs in cases of electromagnetic radiation. This term then leads to the *dipole radiation*. The electric field may be found as

$$\nabla\phi \approx \hat{\mathbf{r}}\frac{1}{c}\frac{\partial\phi}{\partial t} = -\hat{\mathbf{r}}\frac{k}{rc^2}\frac{\partial^2}{\partial t^2}\int(\rho_0 r_s)d^3 r_s. \tag{6.17}$$

The electric dipole radiation field is proportional to the second time derivative of the dipole moment.

For masses the situation is different. $\int(mr_s)d^3 r_s$ finds the location of center of mass of the system. The first time derivative of the center of mass is the momentum of the source, which is a conserved quantity. The second time derivative, the acceleration of the center of mass of the system is going to remain zero. Therefore there is no gravitational dipole radiation. (It can also be said, that if some system appears to be accelerating, then there is another system which is accelerating in the opposite direction, and the gravitational dipole radiations of the two systems will perfectly cancel.)

The third term produces quadrupole radiation. The potential is proportional to the second derivative, and the field is proportional to the third derivative of the quadrupole moment:

$$\nabla\phi \approx \hat{\mathbf{r}}\frac{1}{c}\frac{\partial\phi}{\partial t} = -\hat{\mathbf{r}}\frac{k}{2rc^3}\frac{\partial^3}{\partial t^3}\int(\rho_0 r_s^2)d^3 r_s. \tag{6.18}$$

This term which is much smaller than the dipole radiation term, which is why it may be neglected in a great majority of electromagnetic systems. In cases of electronic transitions in atoms it allows so-called "forbidden transitions."

For gravitational radiation the quadrupole term is the first non-zero term. Any gravitational radiation comes from mass quadrupole moment.

6.4.1 ROTATING BINARY

Previously, we found the expressions for the first few multipole moments of a "static binary." Let us now suppose that this binary rotates in the xy-plane (Fig. 6.2), with angular velocity ω, so

$$\begin{bmatrix} x_1(t) \\ y_1(t) \\ z_1(t) \end{bmatrix} = \begin{bmatrix} r_1\cos(\omega t) \\ r_1\sin(\omega t) \\ 0 \end{bmatrix} = \hat{x}r_1\cos(\omega t) + \hat{y}r_1\sin(\omega t) \tag{6.19}$$

and

$$\begin{bmatrix} x_2(t) \\ y_2(t) \\ z_2(t) \end{bmatrix} = \begin{bmatrix} r_2\cos(\pi + \omega t) \\ r_2\sin(\pi + \omega t) \\ 0 \end{bmatrix} = -\hat{x}r_2\cos(\omega t) - \hat{y}r_2\sin(\omega t). \tag{6.20}$$

Monopole Moment

The first derivative of the monopole moment of a binary is zero, since the total source amount is constant:

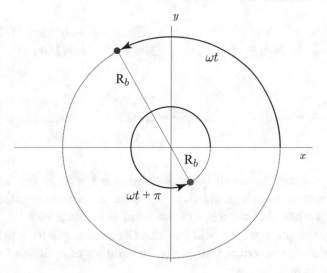

Figure 6.2: Rotating binary.

$$\frac{d}{dt}(S_1 + S_2) = 0. \tag{6.21}$$

Dipole Moment

It should be obvious that the derivatives of a rotating vector are just the rotating vectors.

$$\frac{d}{dt}(S_1 r_1 - S_2 r_2) \begin{bmatrix} \cos(\omega t) \\ \sin(\omega t) \\ 0 \end{bmatrix} = \omega(S_1 r_1 - S_2 r_2) \begin{bmatrix} -\sin(\omega t) \\ \cos(\omega t) \\ 0 \end{bmatrix} \tag{6.22}$$

and similarly for higher derivatives. In case of gravitational interaction, we obtain zero for all orders.

Quadrupole Moment

The first non-trivial result for the gravitational interaction is obtained for the quadrupole.

$$\begin{aligned}
\frac{dQ}{dt} &= \frac{d}{dt}\left(S_1 r_1^2 + S_2 r_2^2\right) \begin{bmatrix} \cos^2 \omega t - \frac{1}{3} & \sin \omega t \cos \omega t & 0 \\ \sin \omega t \cos \omega t & \sin^2 \omega t - \frac{1}{3} & 0 \\ 0 & 0 & -\frac{1}{3} \end{bmatrix} \\
&= \omega\left(S_1 r_1^2 + S_2 r_2^2\right) \begin{bmatrix} \sin(2\omega t) & \cos(2\omega t) & 0 \\ \cos(2\omega t) & \sin(2\omega t) & 0 \\ 0 & 0 & 0 \end{bmatrix},
\end{aligned} \tag{6.23}$$

$$\frac{d^2 Q}{dt^2} = 2\omega^2 (S_1 r_1^2 + S_2 r_2^2) \begin{bmatrix} \cos(2\omega t) & -\sin(2\omega t) & 0 \\ -\sin(2\omega t) & \cos(2\omega t) & 0 \\ 0 & 0 & 0 \end{bmatrix}, \qquad (6.24)$$

$$\frac{d^3 Q}{dt^3} = 4\omega^3 (S_1 r_1^2 + S_2 r_2^2) \begin{bmatrix} -\sin(2\omega t) & -\cos(2\omega t) & 0 \\ -\cos(2\omega t) & -\sin(2\omega t) & 0 \\ 0 & 0 & 0 \end{bmatrix}, \qquad (6.25)$$

and so on.

The frequency of the change of quadrupole moment is *twice* the frequency of rotation of the binary. This means, in particular, that the frequency of the radiation field produced by the rotating binary is twice the frequency of the rotation of the binary. This should not be too surprising if one remembers what the field produced by a static quadrupole looks like.

We may use these results now to find the magnitudes of radiation fields of the rotating binary, as well as the radiated power.

6.5   ESTIMATES FROM THE MULTIPOLE EQUATIONS

6.5.1   RADIATED POWER, OR *LUMINOSITY*

As discussed in the previous chapter, the radiation produced by the retarded "inverse square law" sources, concentrated within a bounded region, may be described in terms of multipoles. The nth multipole is a n-th order tensor, whose elements are of the form $\int \rho r_s^n d^3 r_s$. Any multipole, if it changes with time, produces potential $\phi_n(\mathbf{r})$, which (apart from a complicated angular dependence)falls as $1/r$ with the distance r from the center of mass of the bounded source region.

Let us use dimensional analysis to determine the form of the equations for the power of the produced radiation. This power is called *luminosity* in GW research. We should expect this power to depend on

- the magnitudes of the sources S,

- the typical size of the source r_s,

- the frequency of oscillations within the source ω,

- the speed of propagation of the radiation c,

- the constant k,

Attempts to produce a quantity with units of power are too successful: there is a multitude of formulas that may be produced. The choice may be restricted by realization that for multipoles, instead of considering the magnitude and the typical size of the source, we should be considering the combination, $S r_s^n$. So, the dipole, whose strength is $S r$ is described with $n = 1$, for quadrupole it is $n = 2$, etc. The field strength is proportional to the corresponding multipole moment, and the power to its square:

$$[P] = ([S][r_s]^n)^2 k^\alpha \omega^\beta c^\gamma. \tag{6.26}$$

Now let's examine the units of the individual terms. The obvious ones are

$$[c] = LT^{-1},$$

$$\omega = T^{-1},$$

$$[P] = [M][V]^2[T]^{-1} = [M][L]^2[T]^{-3}.$$

The units of $[S]$ are either the units of charge or of mass. In any case they should cancel out. The units of k are obtained from any of the defining equations, such as the inverse square law:

$$[F] = [k]\frac{[S]^2}{[L]^2} \quad \text{therefore} \quad [k] = ML^3T^{-2}[S]^{-2}. \tag{6.27}$$

Putting all these into 6.26 we have

$$[M][L]^2[T]^{-3} = ([S]^2 L^{2n})(ML^3T^{-2}[S]^{-2})^\alpha T^{-\beta}(LT^{-1})^\gamma$$

yielding

$$\alpha = 1,$$

$$\beta = 2(n + 1),$$

and

$$\gamma = -(2n + 1).$$

Thus, from the dimensional considerations, the power produced by the nth multipole moment of the source is

$$P = A\frac{k(Sr_s^n)^2\omega^{2(n+1)}}{c^{2n+1}}, \tag{6.28}$$

where A is a numerical constant which cannot be found by mere dimensional analysis.

The above formulas produce correct results for electric dipole oscillation

$$P = A\frac{1}{4\pi\epsilon_0}\frac{d^2\omega^4}{c^3},$$

where $d \equiv qr_s$, the dipole moment (the value of constant A is $A = 1/3$ for a linear oscillating dipole, and $A = 2/3$ for a rotating dipole).

For the quadrupole radiation, we obtain

$$P = A\frac{1}{4\pi\epsilon_0}\frac{Q^2\omega^6}{c^5}$$

with the exact value of A dependent on the type of the quadrupole. In particular, for the case of our "binary" with two identical sources traversing the same circle at the two ends of a diameter, the value is $A = 32/5$.

These expressions for radiated power may be used right away to find the rate of decrease of the internal energy of the radiating source. In particular, as will be discussed in Chapter 20 the first observational evidence of the existence of GWs was obtained from the observation of the decrease of rotational energy of an orbiting neutron star binary.

CHAPTER 7

Magnetism and Gravito-Magnetism

7.1 GRAVITO-MAGNETISM HAS TO EXIST

In the previous sections we have found that postulating the finite propagation time for a single potential ϕ would be sufficient to obtain waves. Many of the properties of the waves obtained in such manner agree with the properties of the real EM waves and GW.

The simple generalization of Newtonian gravity to include finite speed propagation has been introduced as far back as year 1805 by Laplace. He used the ideas based on the notion of *astronomical aberration* to show that the speed of propagation of gravitational field has to be much larger than the speed of light (in fact, seven million times as large). The understanding of his reasoning is useful to the understanding of why magnetic-like fields have to exist in any theory in which gravity travels at finite speed.

Astronomical aberration of light is a phenomenon which produces an apparent motion of celestial objects about their true positions, dependent on the velocity of the observer. For instance, the apparent position of the Sun in the sky is displaced from the actual position because it takes light eight minutes to reach the Earth. Suppose now that the force of gravity also propagates at the finite speed. The Earth is then attracted not to the *actual*, but to the apparent position of the Sun. That means that the force of gravity, effectively, stops being the *central force*. Instead, there is now an additional component of the force, proportional to the amount of aberration, v_{Earth}/c, pointing in the direction of Earth's motion. That non-radial component of the force should produce non-radial acceleration, removing the energy from the Sun-Earth system.

This loss of energy is different from the loss of energy due to GW radiation. One glaring difference is in the rate of energy loss: if this loss were real, the Earth-Sun system would collapse within approximately 400 years.

However, as we know now, according to the GR the gravitational fields do, indeed, propagate at the speed of light. How is the apparent paradox resolved?

There is the aberration paradox in electrodynamics as well. Here at issue is not the radiative instability of a classical atom: that conundrum can be resolved only through the application of quantum mechanics. The paradox is different: it is a well-known fact that the electric field of a single moving point charge is directed along the *present* position of the charge, even if the charge is moving at relativistic speeds (e.g., [2], Section 10.3). How can this fact be reconciled with the idea that electric fields propagate at the speed of light?

The resolution of the aberration paradox in electrodynamics comes from the fact that the electric field is not solely produced by the electric potential. There are contributions to the electric field from both the scalar and vector potentials:

$$\mathbf{E} = -\nabla\phi - \frac{\partial \mathbf{A}}{\partial t}. \tag{7.1}$$

The two contributions "contrive" to produce the electric field pointing in the direction to or from the present position of the charge. Of course, the very existence of the vector potential implies the existence of magnetic fields.

The introduction of the equivalent of magnetic field for gravitational forces can resolve Laplace's paradox. It does not provide the full theory of gravity, but is sufficient to describe weak gravitational fields. Once that is done, we may attempt to proceed to use the procedures and the results of electrodynamics to predict the behavior of gravitating masses. Based on the correspondence, we may write a set of "gravimaxwell equations," and use them to introduce GWs by precise analogy.

We will now use these ideas to calculate the signal from the GW emission.

7.2   GRAVITATIOANAL WAVES FROM GRAVITY-ELECTROMAGNETISM ANALOGY

The most convenient approach to EM waves is through electromagnetic potentials: one *scalar potential φ*, which is usually introduced in electrostatic, and one *vector potential* **B** , which, as the name suggests, has three components. The electromagnetic fields **E** and **B** are then obtained from the potentials by taking spatial and time derivatives. In electrostatics, for example, $\mathbf{E} = -\nabla\phi$ will give a conservative non-rotational electrostatic field. In magnetostatics, on the other hand, the prescription $\mathbf{B} = \nabla \times \mathbf{A}$ produces circulating magnetic fields which "wrap around" wires.

These potentials are not unique. For instance, in electrostatics, the same constant number may be added to the electric potential at every point in space. Since the gradient of a constant is zero, the resulting **E** will come out the same.

Even though it is possible to calculate the electrostatic and magnetostatic fields directly from the charges and currents, it is usually much more convenient to calculate the potentials first, then apply the appropriate prescriptions to obtain the fields. For a trivial example, the potential of a single point charge is

$$\phi = k\frac{q}{|\mathbf{r}_s - \mathbf{r}_d|} + \text{const.}$$

The electrostatic field may be found by taking three derivatives, along each coordinate direction. Obtaining electrostatic field directly by using

$$\mathbf{E} = k\frac{q}{|\mathbf{r}_s - \mathbf{r}_d|^2}\hat{\mathbf{r}}$$

instead requires finding projections of vector components directly. This becomes rapidly more involved as the number of charges increases.

To study waves we need to study time-varying fields. Time-varying electric fields may have non-zero circulation. For instance, a magnetic field which is uniform within a region in space, but which is constantly increasing in time, gives rise to a circulating electric field. An electric charge placed in such a field would experience a constant angular acceleration.

All such phenomena are still well described by the electromagnetic potentials, only the prescriptions for obtaining the electromagnetic fields are more complicated.

Just as it is possible to add a constant number to the electrostatic potential, and still obtain the same electrostatic field, it is possible to modify the full electromagnetic potentials and, still obtain the same electromagnetic fields. The different ways of writing pointentials are called *gauges*. Two gauges are commonly used in classical electromagnetis: *Coulomb gauge* and *Lorenz gauge* (many more are possible [13]).

7.2.1   LORENZ GAUGE

The most intuitive gauge is the Lorenz gauge. The potentials are calculated in this gauge by the exact analogies to the static case, except the potentials are defined to propagate in space at the speed of light. Therefore, retarded potentials are used for both scalar and vector potentials:

$$\phi_{Lorenz}(\mathbf{r}_d, t_d) = -k \int \frac{\rho(\mathbf{r}_s, t_d - |\mathbf{r}_s - \mathbf{r}_d|/c)}{|\mathbf{r}_s - \mathbf{r}_d|} d^3 r_s$$

$$\mathbf{A}_{Lorenz}(\mathbf{r}_d, t_d) = -\frac{k}{c^2} \int \frac{\mathbf{J}(\mathbf{r}_s, t_d - |\mathbf{r}_s - \mathbf{r}_d|/c)}{|\mathbf{r}_s - \mathbf{r}_d|} d^3 r_s. \tag{7.2}$$

Just as the quantity ρ stands for charge density in electromagnetism, or for mass density in its gravitational analogy, so the vector $\mathbf{J}$ stands for electric current density in in electromagnetism, and for mass current density in the gravitational analogy. In particular, for a moving point charge, $\mathbf{J} = q\mathbf{v}$, and for a moving point mass $\mathbf{J}$ is simply the linear momentum, $\mathbf{J} = m\mathbf{v}$.

We see that in addition to the scalar potential ϕ, every component of the vector potential $\mathbf{A}$ follows the same equation. The electric and magnetic fields are obtainable by prescription

$$\mathbf{E} = -\nabla\phi_{Lorenz} - \frac{\partial\mathbf{A}_{Lorenz}}{\partial t},$$

$$\mathbf{B} = \nabla \times \mathbf{A}_{Lorenz}. \tag{7.3}$$

The more complicated counterpart to the Coulomb's equation which produces fields directly out of charge and current densities are called either *Schott's formulae* or *Jefimenko equations*.

To calculate the radiation using this additional term we have to repeat the expansion into multipole moments for each component of the new source terms, $\dot{\mathbf{J}}$. We can repeat exactly all the arguments of the previous chapter, for both the density ρ and current density $\dot{J}_i$ for each component of $\mathbf{J}$.

All of the results obtained in the previous chapter had wrong polarization: longitudinal, in the direction from/to the source. The multipole expansion of the vector potential in Lorentz gauge produces three vector components: two transverse, and one longitudinal. The two longitudinal components, one from the scalar potential, the other from vector potential, exactly cancel. What is left are the two transverse components.

7.2.2   COULOMB GAUGE

For radiation problems it turns out to be easier to use the less intuitive Coulomb gauge. In that gauge

$$\phi_{Coulomb}(\mathbf{r}_d, t_d) = -k \int \frac{\rho(\mathbf{r}_s, t_d)}{r} d^3 r_s$$
$$\mathbf{A}_{Coulomb}(\mathbf{r}_d, t_d) = -\frac{k}{c^2} \int \frac{\mathbf{J}(\mathbf{r}_s, t_d - r/c)}{r} d^3 r_s. \tag{7.4}$$

The scalar potential is calculated as if all sources were at rest, which is, obviously, easier. However the biggest convenience associated with this gauge becomes apparent when we look at the prescription for calculating the electric (or gravielectric) fields:

$$\mathbf{E} = -\frac{\partial \mathbf{A}_{transverse}}{\partial t}. \tag{7.5}$$

The scalar potential does not enter this formula at all. In Coulomb gauge the radiation field depends only on the transverse components of the vector potential. Instead of calculating four components (1 from scalar and 3 from vector potential) to have two of them cancel each other, we just drop the longitudinal component of vector potential at the outset.

We will use this method to obtain the results for the radiation of the binary.

7.3   ROTATING BINARY

For a point source S, the source current density is simply: $\mathbf{J} = S\mathbf{v}$, where $\mathbf{v}$ is the velocity of the source. For a rotating binary, the coordinates are given by Eqs. (6.19) and (6.20), so the velocities are

$$\mathbf{v}_1 = r_1\omega \begin{bmatrix} -\sin(\omega t) \\ \cos(\omega t) \\ 0 \end{bmatrix} \qquad \mathbf{v}_2 = -r_2\omega \begin{bmatrix} -\sin(\omega t) \\ \cos(\omega t) \\ 0 \end{bmatrix}. \tag{7.6}$$

The vector potential at the detector point is

$$\mathbf{A}(\mathbf{r}_d, t_d) = k \frac{S_1 \mathbf{v}_1(t_1)}{|\mathbf{r}_1(t_1) - \mathbf{r}_d|} + k \frac{S_2 \mathbf{v}_2(t_2)}{|\mathbf{r}_2(t_2) - \mathbf{r}_d|}. \tag{7.7}$$

We could now expand the denominators using binomial theorem, however that would result in a series of terms falling as increasing powers of r_d as we have seen in Chapter 5. Instead, we keep only the first term in the expansion of the denominators:

$$\mathbf{A}(\mathbf{r}_d, t_d) = k \frac{S_1 \mathbf{v}_1(t_1)}{r_d} + k \frac{S_2 \mathbf{v}_2(t_2)}{r_d} \tag{7.8}$$

and work on expanding the numerators in time, as was done in Chapter 6: the times t_1 and t_2 at which the potentials were produced are expressed in terms of time t_d when these potentials arrive at the detector:

$$\mathbf{A}(\mathbf{r}_d, t_d) = k \frac{S_1 \mathbf{v}_1(t_d - r_1/c)}{r_d} + k \frac{S_2 \mathbf{v}_2(t_d - r_2/c)}{r_d}. \tag{7.9}$$

This contains two velocities which depend on two different retarded times: $\mathbf{v}_1(t_d - r_1/c)$ and $\mathbf{v}_2(t_d - r_2/c)$. We define a single retarded time $t_d - r_d/c$ from the origin (where the center of the binary's orbits is located). Recalling that $|\mathbf{r}_d - \mathbf{r}_i| = \hat{\mathbf{r}}_d \cdot \mathbf{r}_i$, we can expand the behavior of each of the S_i in terms of the time difference

$$\mathbf{v}_i(\mathbf{r}_i, t_d - r/c) = \mathbf{v}_i(\mathbf{r}_i, t_d - r_d/c) + \frac{r_s}{c} \frac{\partial \mathbf{v}(\mathbf{r}_i, t_d - r_d/c)}{\partial t}$$
$$+ \frac{1}{2!} (\frac{r_1}{c})^2 \frac{\partial^2 \mathbf{v}_i(\mathbf{r}_i, t_d - r_d/c)}{\partial t^2} + \dots. \tag{7.10}$$

We will not use the second or any higher derivative terms. The first derivative of velocity is centripetal acceleration:

$$\frac{d\mathbf{v}_i}{dt} = -\omega^2 \mathbf{r}_i, \tag{7.11}$$

so the time expansion of the velocity simplifies to

$$\mathbf{v}_i(\mathbf{r}_i, t_d - r/c) = \mathbf{v}_i(\mathbf{r}_i, t_d - r_d/c) - \frac{r_s}{c} \omega^2 \mathbf{r}_i + \dots. \tag{7.12}$$

Putting this back into the expression for the vector potential Eq. (7.9),

$$\mathbf{A}(\mathbf{r}_d, t_d) = \frac{k}{c^2 r_d} [S_1 \mathbf{v}_1(t_0) + S_2 \mathbf{v}_2(t_0)]$$
$$- \frac{k\omega^2}{c^3 r_d} [S_1 \mathbf{r}_1(\hat{\mathbf{r}}_d \cdot \mathbf{r}_1) + S_2 \mathbf{r}_2(\hat{\mathbf{r}}_d \cdot \mathbf{r}_2)]. \tag{7.13}$$

The first term, $\frac{k}{c^2 r_d}[S_1 \mathbf{v}_1(t_0) + S_2 \mathbf{v}_2(t_0)]$, is equivalent to the time derivative of the dipole moment in Eq. (6.22). As discussed in the previous chapter, it produces EM dipole radiation if S_i represent electric charges. If, however, S_i represent masses m_i, the quantity

$$m_1 \mathbf{v}_1(t_0) + m_2 \mathbf{v}_2(t_0) = \text{const}$$

by momentum conservation, so this term cannot produce GWs.

The second term contains combinations of terms whose magnitude $S r_s^2$ correspond to the quadrupole moments. Inserting the values of $\mathbf{r}_1$ and $\mathbf{r}_2$ from Eq. (6.19) produces

$$\mathbf{A}(\mathbf{r}_d, t_d) = -\frac{k\omega^2}{c^3 r_d}[S_1 \mathbf{r}_1 (\hat{\mathbf{r}}_d \cdot \mathbf{r}_1) + S_2 \mathbf{r}_2 (\hat{\mathbf{r}}_d \cdot \mathbf{r}_2)] \tag{7.14}$$

$$= -\frac{k\omega^2}{c^3 r_d}(S_1 r_1^2 + S_2 r_2^2) \begin{bmatrix} \cos(\omega t) \\ \sin(\omega t) \\ 0 \end{bmatrix} [r_{dx} \cos(\omega t) + r_{dy} \sin(\omega t)] \tag{7.15}$$

$$= -\frac{k\omega^2}{2c^3 r_d}(S_1 r_1^2 + S_2 r_2^2) \begin{bmatrix} r_{dx}[1 + \cos(2\omega t)] + r_{dy} \sin(2\omega t) \\ r_{dx} \sin(2\omega t) + r_{dy}[1 + \cos(2\omega t)] \\ 0 \end{bmatrix}. \tag{7.16}$$

We are ready to use the prescription for obtaining the fields out of magnetic potential in Eq. (7.5). Taking the time derivative yields

$$\mathbf{E} = -\frac{d}{dt}\mathbf{A}(\mathbf{r}_d, t_d)$$

$$= \frac{k\omega^3}{c^3 r_d}(S_1 r_1^2 + S_2 r_2^2) \begin{bmatrix} r_{dx} \sin(2\omega t)] - r_{dy} \cos(2\omega t) \\ -r_{dx} \cos(2\omega t) + r_{dy} \sin(2\omega t) \\ 0 \end{bmatrix}. \tag{7.17}$$

We have obtained more rigorously the results of the previous sections: the field

- falls inversely proportionally with the distance r_d

- is determined by the quadrupole moment $Q = S_1 r_1^2 + S_1 r_1^2$

- oscillates at twice the frequency of rotation 2ω.

7.3.1   RADIATED POWER

The radiative field is carries energy away from the source. In EM theory, the *intensity* of a wave is the average power per unit area transported by the wave. Its value is given by

$$u = \frac{1}{2}\left(c\epsilon_0 E^2\right).$$

We can generalize this definition as

$$u = \frac{c}{4\pi k} < E^2 >,$$

(7.18)

where the angle brackets $< \ldots >$ denote the time average.

CHAPTER 8

Tidal Field as Gravielectric Field

The equivalence principle implies that the forces produced by gravitational fields are locally indistinguishable from pseudo-forces produced by acceleration. The difference between the two situations in the *gradients* of those forces. A linearly accelerating observer finds a constant pseudo-force acting on any object independent of its position. If the observed force changes from location to location within the coordinate system it may be inferred that something other than linear acceleration is taking place.

It makes sense then, that instead of measuring the forces produced by gravitational fields, we should be measuring how those forces vary from one location to another. In other words we should be measuring the gradient of the forces. This may be conveniently done by comparing the accelerations of at least two free-falling test masses (TMs) at different spatial locations. The relative motion of two free-falling TMs should inform the observer whether there is a localized source of gravity nearby.

If one free falling TM happens to be placed closer to the gravitating source, it will experience a larger acceleration than the TM located farther away. The two TMs located along the radial direction from the source will separate further from each other. On the other hand, if the two TMs happen to be initially placed at the same distance from the CM of the source, they will both be falling toward the CM, thereby approaching each other. This process, commonly known as "spaghettification," is due to the *tidal field* (Fig. 8.1).

To make this more precise, it is more convenient to abandon the discussion of gravitational or pseudo-forces, and to talk directly in terms of accelerations of the TMs. So, we consider the "gravitational acceleration field" $\mathbf{a(r)}$.

Let the two TMs be separated by a small χ, the *separation vector*. In the presence of gravity the two accelerations-due-to-gravity will be different, and the two masses will have a component of acceleration toward (or away from) *each other*. This is the *tidal acceleration*:

$$a_{tidal}(1,2) = \mathbf{a(r_1)} - \mathbf{a(r_2)} = -\chi \cdot \nabla \mathbf{a(r)}, \qquad (8.1)$$

where we have expanded the difference in a Taylor series right away.

The acceleration field itself is a vector field, and $\nabla \mathbf{a}$ is a 2nd rank tensor field. We identify it with the static *gravielectric field*:

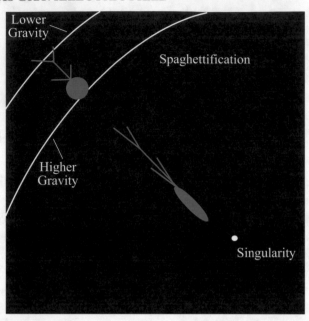

Figure 8.1: Spaghettification of a hapless explorer falling into a black hole. The explorer is both stretched and compressed: stretched in the radial, and compressed in the transverse directions.

$$\mathcal{E}(\mathbf{r}) \equiv -\nabla \mathbf{a}(\mathbf{r}).$$ (8.2)

This 2nd rank tensor may be represented by a matrix in various coordinate systems, e.g., in Cartesian coordinates

$$\mathcal{E} = \begin{bmatrix} \frac{\partial a_x}{\partial x} & \frac{\partial a_x}{\partial y} & \frac{\partial a_x}{\partial z} \\ \frac{\partial a_y}{\partial x} & \frac{\partial a_y}{\partial y} & \frac{\partial a_y}{\partial z} \\ \frac{\partial a_z}{\partial x} & \frac{\partial a_z}{\partial y} & \frac{\partial a_z}{\partial z} \end{bmatrix}.$$ (8.3)

Given two TMs, one at $\mathbf{r}$, the other a very short distance away, we can use the value of $\mathcal{E}(\mathbf{r})$ to calculate how the initial separation of two masses along the direction, say, x results in their relative acceleration in the direction, say, z. More generally, the tidal acceleration between two TMs at $\mathbf{r}$ separated by small $\boldsymbol{\chi}$ is

$$\mathbf{a}_{tidal}(1,2) = -\boldsymbol{\chi} \cdot \mathbf{E}(\mathbf{r}).$$ (8.4)

As an example, outside a weakly gravitating spherical mass (e.g., Earth), the vector of Newtonian gravitational acceleration is

$$\mathbf{a}_N = -\hat{\mathbf{r}}\frac{GM}{r^2}.$$
(8.5)

Applying the definition Eq. (8.2) to this potential, obtain in Cartesian and in radial coordinates

$$\mathcal{E} = \frac{GM}{r^5}\begin{bmatrix} 3x^2 - r^2 & 3xy & 3xz \\ 3yx & 3y^2 - r^2 & 3yx \\ 3zx & 3zy & 3z^2 - r^2 \end{bmatrix} = \frac{GM}{r^3}\begin{bmatrix} 2 & 0 & 0 \\ 0 & -1 & 0 \\ 0 & 0 & -1 \end{bmatrix}.$$

Several features of these tensors should be noted. There is an obvious similarity of the acceleration tensor of a point source above to a part of the quadrupole tensor. Comparing them side by side makes this obvious:

$$Q_{ij} \equiv \int [3r_{si}r_{sj} - r_s^2\delta_{ij}]\rho(\mathbf{r}_s)d^3r$$

$$\mathcal{E} = [3r_ir_j - r^2\delta_{ij}]\frac{GM}{r^5}.$$

This is not surprising, since both object appear as a result of taking the second derivatives of the same point source potential. In fact, the potential of an arbitrarily shaped object can be built out of the potentials of point sources, bringing the analogy even closer.

The great usefulness of gravielectric tidal fields is that the analogue of Maxwell equations for gravity is better created using these tensors (along with the gravimagnetic frame-dragging tensors of the next chapter), than with the force-derived vector fields of the previous chapters. The tidal field $\mathcal{E}$ can directly describe the effect that the gravitating source has on the *pairs* of test masses. The gravielectric tensors are symmetric (e.g., $\mathcal{E}_{xy} = \mathcal{E}_{yx}$). The trace of a gravielectric tensor can be non-zero only at the location of a gravitating mass:

$$\Sigma_k \mathcal{E}_{kk} = \nabla^2 \Phi.$$

CHAPTER 9

Frame-Drag Field as Gravimagnetic Field

As was discussed in Chapter 7, because of 4-dimensional nature of spacetime, moving sources produce additional fields. Having established that the proper analogue of vector electric fields are rank-2 tensor gravielectric fields, rather than gravitational force fields, we now look at the rank-2 tensor gravimagnetic fields.

The prototypical detector of a static magnetic field is a current loop. The gravimagnetic field has a similar relationship with rotating masses. We introduce a *test gyro*: a small *spinning* test mass. The motion of a large gravitating mass will change the direction of the test gyro. The gravimagnetic field describes this change in direction.

In the local inertial frame of the gyro it has angular momentum **s**. In the absence of any gravitational fields the angular momentum of a stationary test gyro will continue pointing in the same direction: $\mathbf{s}(t) -$ const. If the direction of the angular momentum is changing we talk about *precession*: the slow rotation of the direction of the angular momentum vector.

The test gyro may, actually, precess due to several effects. If the test gyro is moves around a circe with speed β it will precess even in the absence of any gravitational fields (e.g., if the centripetal force is provided by electromagnetic forces of attraction). This is the *Thomas precession*. Thomas precession is a special-relativistic effect, resulting from the fact that the moving gyro undergoes special-relativistic time dilation and length-contraction along the direction of its instantaneous velocity. As a result, after a full revolution its rest-frame is tilted with respect to its original orientation by angle

$$\theta_{Thomas} = -2\pi(1 - \frac{1}{\sqrt{1 - \beta^2}}) \approx \pi\beta^2.$$

The other two precessions follow from the GR, but only should be described as due to gravimagnetic fields. The *geodetic precession* is evident whenever the test gyro is moved around the gravitating mass, irrespective of the rotation of the gravitating mass itself. If the test gyro rotates in a closed orbit around a gravitating mass, e.g., a planet or a star, its angular momentum experiences geodetic precession by angle

$$\Delta\phi_{geodetic} = 2\pi \left[1 - \sqrt{1 - \frac{3GM}{c^2 R}} \right] \quad \text{(per orbit)}.$$

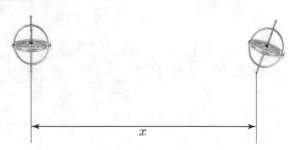

Figure 9.1: A pair of test gyros.

The existence of this precession has been verified experimentally since 2007, when the Gravity Probe B satellite measured the precession of a test gyro in a circular orbit around the Earth [20]. The magnitude of this precession is on the order of $8.4''(R_E/R)^{5/2}$ per year (where R is the radius of circular orbit of the satellite).

Finally, *Lense–Thirring precession* of the test gyro appears only if the gravitating mass itself is rotating. This is the precession that can be described as being due to gravimagnetic fields. In a direct analogy to the generation of a magnetic field by a solenoid, a large rotating mass (e.g., a single star rotating around its own axis) generates gravimagnetic field. Just as a magnetic field set up by a solenoid may be detected by the torque produced upon a test current loop, the gravimagnetic field is evident in its effect upon the test gyro. Specifically, if the star possesses angular momentum **J**, the test gyro located distance r away from its center experiences the precession of frequency

$$\omega_{LT} = \frac{G}{c^2 r^3}[3(\mathbf{J} \cdot \hat{\mathbf{r}})\hat{\mathbf{r}} - \mathbf{J}]. \tag{9.1}$$

Thus, the angular momentum **J** of the rotating mass plays the role of gravimagnetic moment.

Just as it was not sufficient to consider a single test mass to establish the presence of a gravitational field, it is necessary to consider *two* test gyros to describe the gravimagnetic field.

To establish the 2nd rank tensor gravimagnetic field B_{jk} we need to examine the *differential* precession of two test gyros separated by a small separation vector χ. In the presence of gravity the two precessions-due-to-gravity will be different, and the two gyros will precess toward (or away from) *each other*. This is the differential precession. Viewed from the coordinate system connected to one of the gyros, the other gyro changes only the direction (and not the magnitude) of its angular momentum s. Therefore the rate of change of s can be written as

$$\dot{\mathbf{s}} = \boldsymbol{\Omega} \times \mathbf{s}. \tag{9.2}$$

The $\boldsymbol{\Omega}$ is the vector of *frame dragging angular velocity*. This vector is defined at each point in space around a moving gravitating source, constituting a vector field. As was done with the gravielectricity, we identify the static *gravimagnetic field* with the 2nd order tensor obtained by taking the gradient:

$$\mathcal{B}(\mathbf{r}) \equiv \nabla \mathbf{\Omega}(\mathbf{r}). \tag{9.3}$$

Given two test gyros initially oriented in the same direction, one at $\mathbf{r}$, the other a very short distance away, we can use the value of $\mathcal{B}(\mathbf{r})$ to calculate how the initial separation of two gyros along the direction, say, x results in them being oriented in different directions.

$$\mathbf{\Omega}_{fd}(1, 2) = \boldsymbol{\chi} \cdot \mathcal{B}(\mathbf{r}). \tag{9.4}$$

As an example, outside a *rotating* weakly gravitating spherical mass (e.g., Earth), the components of the frame-drag field are

$$\mathcal{B}_{rr} = -2\mathcal{B}_{\theta\theta} = -2\mathcal{B}_{\phi\phi} = -\frac{6S \cos\theta}{r^4} \tag{9.5}$$

$$\mathcal{B}_{r\theta} = \mathcal{B}_{\theta r} = -\frac{3S \sin\theta}{r^4}. \tag{9.6}$$

The gravimagnetic field affects the relative acceleration of two gravitating test masses. The full expression is very reminiscent of the Lorentz force equation for EM forces:

$$\mathbf{a}(1, 2) = -\boldsymbol{\chi} \cdot (\mathcal{E} + \mathbf{v} \times \mathcal{B} - \mathcal{B} \times \mathbf{v}). \tag{9.7}$$

CHAPTER 10

Gravitational Waves in Terms of Tidal and Frame-Drag Fields

The tidal and frame-drag fields are derivable as approximations from GR. In the empty space tensors $\mathcal{E}$ and $\mathcal{B}$ are symmetric and traceless, and so have ten independent components. In the same approximation, the field equations for these fields strongly resemble Maxwell equations. In empty space they take on form

$$
\begin{aligned}
\nabla \cdot \mathcal{E} &= 0 \\
\nabla \cdot \mathcal{B} &= 0 \\
\nabla \times \mathcal{B} &= \frac{1}{c}\frac{\partial \mathcal{E}}{\partial t} \\
\nabla \times \mathcal{E} &= -\frac{1}{c}\frac{\partial \mathcal{B}}{\partial t}.
\end{aligned}
\tag{10.1}
$$

In these *gravimaxwell equations* divergence and curl act on the 2nd rank tensors. The interpretation of these operations is analogous to the interpretations these operations have when applied to vector fields.

So, the divergence of a vector field is a scalar field. At each point in space the value of this scalar field represents the flux of the original vector field out of a small volume around that point. The divergence of a 2nd rank tensor field is a vector field. Each component of this vector of divergence represent the flux of the portion of tensor field out of a small volume around each point in space. Similarly, the curl of a vector field is another vector field, which represents the "circulation" of the original field. The curl of a tensor field is a tensor field, which represents the "circulation" of the components of the original tensor field.

Just as ordinary Maxwell equations can be satisfied by plain wave of the form $\mathbf{E}(\mathbf{r}, t) = \mathbf{E}\cos(\mathbf{k}\cdot\mathbf{x} - \omega t)$, with a corresponding expression of $\mathbf{B}$, the gravielectric and gravimagnetic fields can be satisfied by plain waves of the form

$$\mathcal{E}_{ij}(x, y, z, t) = \frac{1}{2}\mathcal{E}_{ij}^0 \cos(\mathbf{k} \cdot \mathbf{x} - \omega t)$$
$$\mathcal{B}_{ij}(x, y, z, t) = \frac{1}{2}\mathcal{B}_{ij}^0 \cos(\mathbf{k} \cdot \mathbf{x} - \omega t), \tag{10.2}$$

where $\mathcal{E}_{ij}^0$ and $\mathcal{B}_{ij}^0$ are constant amplitude tensors. To satisfy the gravimaxwell equations these waves have to be transverse: this is checked in exactly the same way as is done with ordinary Maxwell.

CHAPTER 11

Overview of General Relativity

Electromagnetism has not one, but four potentials: the scalar potential Φ, and the vector potential $\mathbf{A}$. These give rise to two vector fields, $\mathbf{E}$ and $\mathbf{B}$, which constitute the six elements of a second-rank, antisymmetric field-strength tensor.

GR takes the component of the metric tensor as multiple gravitational field quantities. It may be said that the theory possesses in not one, but ten "gravitational potentials"—the ten metric coefficients $g_{\mu\nu}$.

Electromagnetism is a linear theory. The principle of superposition ensures that the potentials and field components may be added using basic laws of scalar and vector addition.

GR, on the other hand, was designed almost from the outset to be a non-linear theory, to ensure that the energy of the gravitational field also acts as its own source. If the theory of gravity were linear then we could, in principle, extract an infinite amount of energy from arbitrary electrically neutral masses. Gravitational sources possess only one "sign," and gravitational potential energy is given by the $1/r$ term. Therefore, if the theory is not modified, simply bringing the masses closer together permits extraction of ever greater amount of energy. The nonlinear terms in GR limit the amount of the total energy that can be extracted to a value of the order Σmc^2, the rest mass energy of the masses.

We briefly outline some elements of the GR, just to give the idea of the connection between geometry and fields.

GR is based on the extension of the principle of relativity: "Identical experiments carried out in different inertial frames produce identical results" to non-inertial systems. Preparing a review article on Special Relativity, Einstein asked himself how the Newtonian theory of gravitation would have to be modified to abide by the relativistic laws.

"Then there occurred to me the...happiest thought of my life...The gravitational field has only a relative existence.... *Because for an observer falling freely from the roof of a house there exists—* at least in his immediate surroundings—*no gravitational field*. Indeed, if the observer drops some bodies then these remain relative to him in a state of rest or of uniform motion independent of their particular chemical or physical nature. The observer therefore has the right to interpret his state as 'at rest'...and his environment as field-free relative to gravitation.

The experimentally known matter independence of the acceleration of fall is therefore a powerful argument for the fact that the relativity postulate has to be extended to coordinate systems which, relative to each other, are in non-uniform motion" [7].

To extend relativity principle, we examine the paths and timings of the light rays in *accelerated* systems similar to the way it is done in basic SR. Examining the times of arrival we almost immediately [7–9] are led to

$$\Delta \tau_B = \Delta \tau_A \left(1 + \frac{a \Delta z}{c^2} \right) \tag{11.1}$$

for the time differences between two identical events as reported by two observers located at a distance Δz away from each other along the line of their acceleration.

The equivalence principle then enables us to postulate the same equation for the time perceived by two observers in a gravitational field of field strength g:

$$\Delta \tau_B = \Delta \tau_A \left(1 + \frac{g \Delta z}{c^2} \right). \tag{11.2}$$

Turning this argument around we relate the gravitational field to this time difference. We identify the gravitational potential (of a weak gravitational field)

$$\Delta \phi = g \Delta z \tag{11.3}$$

to the first order, so the gravitational potential causes the *curvature* in the time coordinate:

$$\left(1 + \frac{\phi}{c^2} \right) dt. \tag{11.4}$$

The metric of spacetime with a weak static gravitational field is then

$$\mathbf{ds}^2 = - \left(1 + \frac{\phi}{c^2} \right)^2 c^2 dt^2 + dx^2 + dy^2 + dz^2. \tag{11.5}$$

It should be pointed out that a popular image of gravity as that of curved *space* is actually quite incorrect. As the above discussion shows, the effect of the classical weak static gravitational field on the slowly moving (with respect to the gravitating mass) matter can be described by curvature in the time coordinate of space time.

Two generalizations to the above equation are then necessary. First, we generalize to stronger fields by

$$\left(1 + \frac{\phi}{c^2} \right) \rightarrow e^{\phi/c^2}.$$

The metric of every *static* gravitational field can be expressed as

$$\mathbf{ds}^2 = -e^{2\phi/c^2} c^2 dt^2 + dx^2 + dy^2 + dz^2. \tag{11.6}$$

To be able to describe dynamic gravitational fields, such as fields produced by moving bodies, the second, and drastic, generalization is needed. It becomes necessary to use full spacetime curvature, rather than just time curvature. Then the general metric of spacetime contains every possible combination of the second power of the coordinate increments (that is, every term of the form dx^2 or $dxdy$:

$$\mathbf{ds}^2 = -e^{2\phi/c^2}dt^2 + 2Adtdx + 2Bdtdy + 2Cdtdz$$
$$+ Ddx^2 + 2Edxdy + 2Fdxdz + Gdy^2 + 2Hdydz + Idz^2, \tag{11.7}$$

or, in matrix notation,

$$\mathbf{ds}^2 = [dt \ \ dx \ \ dy \ \ dz] \begin{bmatrix} e^{2\phi/c^2} & A & B & C \\ A & D & E & F \\ B & E & G & H \\ C & F & H & I \end{bmatrix} \begin{bmatrix} dt \\ dx \\ dy \\ dz \end{bmatrix}. \tag{11.8}$$

The nine metric coefficients (A, B, etc.) together with ϕ constitute the ten potentials of GR.

Much physical intuition may be obtained by "completing the square" for all terms that have dt. Then the metric is put into the *canonical form*

$$\mathbf{ds}^2 = -e^{2\phi/c^2}\left[cdt - \frac{1}{c^2}(w_x dx + w_y dy + w_z dz)\right]^2$$
$$+ k_{xx}dx^2 + k_{yy}dy^2 + k_{zz}dz^2 \tag{11.9}$$

with $w_x = -(c/2)Ae^{-2\phi/c^2}$, etc. The k_{ii} coefficients describe the *spatial* curvature. Spacial curvature significantly influences the motion of fast (relative to the source) test particles, and has a large contribution to the trajectory of light. In particular, it is responsible for one-half of the bending of the light around the sun.

The coefficients w_i, on the other hand, are analogous to the vector potential $\mathbf{A}$ of Maxwell equations of electrodynamics. If we evaluate the motion of a test particle of mass m in space with metric described by

$$\mathbf{ds}^2 = -e^{2\phi/c^2}\left[cdt - \frac{1}{c^2}(w_x dx + w_y dy + w_z dz)\right]^2,$$

we find that that it moves like a particle subject to a gravitational version of Lorentz-type law

$$\mathbf{F} = -\nabla\phi + \mathbf{v} \times (\nabla \times \mathbf{w}).$$

Therefore $\mathbf{w}$ is the *gravimagnetic potential* of the gravitational field.

To be able to work with 4-dimensional curved surfaces requires mathematical apparatus of differential geometry and tensor analysis. It is not the goal of this overview to present the full theory. Instead we give a brief outline of the main equations and of an approximation which is valid for weak GWs.

CHAPTER 12

Gravitational Waves and Einstein Equations

As was discussed in the previous charges, weak gravitational fields can be described by Maxwell-like equations, with the gravielectric fields manifesting through tidal, and gravimagnetic fields manifesting through frame dragging effects.

The Einstein equations elaborate on this analogy. Being valid for strong field regimes as well as for weak fields, they introduce non-linear effects. However, full Einstein equations do not describe gravity in terms of fields. Instead, gravity is described by spacetime curvature. In certain situations (specifically, when the spacetime curvature is sufficiently small) the gravielectric and gravimagnetic fields come out as various parts of the curvature tensor.

In full theory the mass-energy of matter is the source of spacetime curvature. The general form of the Einstein equations is

$$\text{local spacetime curvature} = \text{matter-energy density}. \tag{12.1}$$

12.1 SPACETIME CURVATURE

The curvature is what creates the tidal forces on the free-falling particles. The curvature may be characterized by these tidal forces. As was discussed earlier, in weak fields,

$$\frac{d^2\chi^i}{dt^2} = -\delta^{ij}\frac{\partial^2\Phi}{\partial x^j \partial x^k}\chi^k, \tag{12.2}$$

where Φ is the Newtonian gravitational potential, and χ^k is the 3-vector of spatial separation between two test masses.

Full theory generalizes this *Newtonian deviation equation* to a four-dimensional *equation of geodesic deviation*. All operations (such as taking a derivative) are now taken in a curved space-time. In curved coordinates the separation between two points may be changing even if these two points move along *parallel* lines, provided the space between them has a particular curvature. The simplest classic example is the motion of points on the surface of a sphere: two points moving along two parallel meridians will inexorably approach each other as they move toward the pole. To describe this effect, differential geometry uses the concept of *covariant derivatives*. This concept recognizes that two quantities at *different* points of a curved spacetime must not be directly compared, unless they are first brought to the *same* point. Any of the well-known vector operations (vector addition, subtraction, etc.) which are performed on a pair of vectors located at different points in a flat spacetime (e.g., the comparison of the velocities of two spaceships), have to be modified in a curved spacetime.

The two separated points in the curved spacetime are first brought together in a process of *parallel transport*, where one of the vectors is transported parallel to itself to the location of the other vector. This additional operation permits then the definition of *covariant derivative*: the derivative of a vector field at point in a Φ curved spacetime. The covariant derivative can be expressed through standard derivatives by the use of a complicated function called the *Christoffel symbol* $\Gamma^{\gamma}_{\alpha\beta}$:

$$\nabla_\beta v^\alpha = \frac{\partial v^\alpha}{\partial x^\beta} + \Gamma^\alpha_{\beta\gamma} v^\gamma. \tag{12.3}$$

The tidal acceleration of two points is now obtained as the second *covariant* derivative of the spacing between the points. This operation results in the *equation of geodesic deviation*. In the most appropriate coordinate system, which is the *freely falling frame*, the equation of geodesic deviation takes on the form

$$\frac{d^2 \chi^{\hat{\alpha}}}{d\tau^2} = -R^{\hat{\alpha}}_{\hat{\tau}\hat{\beta}\hat{\tau}} \chi^{\hat{\beta}}, \tag{12.4}$$

where $R^{\alpha}_{\beta\gamma\delta}$ is a function of four sets of coordinates called *Riemann curvature tensor*. As the name implies, from the geometrical point of view it describes the curvature of the spacetime. From the more utilitarian point of view, however, Riemann tensor is just a set of several complicated functions of the three spatial and one time coordinate. Collectively, these functions describe the gravitational fields. This is a direct generalization of the Newtonian deviation equation (12.2), where the set of several functions of three spatial variables (obtained by taking the second derivatives of the Newtonian gravitational potential) is replaced by the complicated structure of the Riemann curvature tensor. In the case of weak gravitational fields produced by slowly moving objects, most of these functions simplify to zero, and those that remain reduce to the Newtonian $\frac{\partial^2 \Phi}{\partial x^j \partial x^k}$. However, the parts of this tensor which contain the derivatives with respect to the time coordinate also describe the effects of the retardation in the propagation of the gravitational field, and the gravitoelectric and gravitomagnetic effects discussed earlier. The study of the properties of this tensor in various physical situations constitute a significant part of the study of GR.

12.2   MATTER-ENERGY DENSITY

Newton's law of gravity postulates that mass is the source of gravitational forces, that is, the source of the spatial derivatives of gravitational potential:

$$\nabla^2 \Phi = 4\pi G m. \tag{12.5}$$

Noticing that

$$\nabla^2 \Phi = \delta^{ij} \frac{\partial^2 \Phi}{\partial x^j \, \partial x^k}$$

makes it obvious that Newton's law of gravity can easily be expressed in terms of Newtonian deviation:

$$\delta^{ij} \frac{\partial^2 \Phi}{\partial x^j \, \partial x^k} = 4\pi G m. \tag{12.6}$$

The replacement of the Newtonian deviation with general-relativistic expression involving curvature tensors necessitates the generalization of the source term. The situation may be compared to the transition from electrostatic to electrodynamics. The transition from Gauss's law to the Maxwell's equations is equivalent to the transition from a single electric potential V to the current density 4-vector

$$J^\mu = (c\rho_0, J_x, J_y, J_z) = \frac{\rho_0}{\sqrt{1 - v^2/c^2}} (c, v_x, v_y, v_z), \tag{12.7}$$

where ρ_0 is the charge density in the rest system of the charge. The Maxwell equations describe the relationship between the source of EM fields on one hand, and the spatial and time derivatives of those fields on the other.

Similarly, Einstein equations describe the relationships between the sources of the gravitational fields on one hand, and the curvature tensor on the other. The spacetime metric $g_{\alpha\beta}$ plays the role analogous to the one played by the electromagnetic potentials. The curvature tensor, which involves double derivatives of the metric, plays the role analogous to the that of the derivatives of the electric and magnetic fields in electrodynamics.

It might appear that a quantity analogous to the electric charge-current density should be composed for gravity: after all non-relativistic physics contains the notions of mass density and mass currents. Special relativity defines the energy-momentum 4-vector

$$p^\mu = \frac{m_0}{\sqrt{1 - v^2/c^2}} (c, v_x, v_y, v_z), \tag{12.8}$$

where m is the rest mass of a particle.

However, Einstein realized that the energy-momentum 4-vector cannot be the sole the source of gravitational field. Rather, the source is the energy and momentum density. When a scalar quantity, such as electric charge (or, for that matter, the rest mass), is associated with a volume, it gives rise to a density-current 4-vector (e.g., charge current-density, or energy-momentum). But when a 4-vector is associated with a volume, it gives rise to a second-rank tensor. The source of gravitational fields is such a second-rank tensor, the *energy-momentum-stress tensor*. The components of this tensor are

$$T^{\mu\nu} = \begin{bmatrix} \text{energy density} & \frac{1}{c}\text{energy flux} \\ \frac{1}{c}\text{momentum density} & \text{stress tensor} \end{bmatrix}. \tag{12.9}$$

Putting this all together we obtain, finally, the *Einstein Equation*:

$$R_{\mu\nu} - \frac{1}{2}g_{\mu\nu}R = \frac{8\pi G}{c^4}T_{\mu\nu}. \tag{12.10}$$

The left-hand side contains a complicated combination of second-order derivatives of the components of metric tensor. On the right-hand side there is the energy-momentum-stress source.

Let us now look at the solutions of the Einstein equation. There are no exact solutions of the full Einstein equations that would show the GWs. Until the development of modern computing the work in GR approached problems from broadly the following two directions: either the search for approximate solutions in various approximations, or the mathematical proofs of existence of solutions with various boundary conditions. The *numeric relativity*, that is the numerical search for solutions of various cases of interest has only started to be realizable in the 21st century, and is a lively field of current active research. One of the many exciting advancements made possible by the direct detection of GW is the observational study of generation of GWs.

12.3   GWS IN GR: WEAK FIELD APPROXIMATION

Having said that the relativistic theory of gravity has to be a non-linear theory, we construct a *linearized* theory: a relativistic linear approximation to GR. This approximation is valid when the gravitational fields are weak. Then the energy associated with the gravity is sufficiently small that the non-linearity caused by the gravity being its own source is approximately zero.

Linearization starts with the Lorentzian metric which describes spacetime in the absence of gravity: the flat spacetime

$$ds^2 = -c^2dt^2 + dx^2 + dy^2 + dz^2 \equiv \eta_{\mu\nu}dx^\mu dx^{nu}, \tag{12.11}$$

where

$$\eta_{\mu\nu} \equiv \begin{bmatrix} -1 & 0 & 0 & 0 \\ 0 & 1 & 0 & 0 \\ 0 & 0 & 1 & 0 \\ 0 & 0 & 0 & 1 \end{bmatrix} \tag{12.12}$$

is the Minkowski metric.

In a weak gravitational field the spacetime is "nearly" flat. Then there exists a set of *nearly Lorentz coordinates* in which the metric is Minkowski metric with a small *metric perturbation*:

$$g_{\mu\nu} = \eta_{\mu\nu} + h_{\mu\nu}, \tag{12.13}$$

where $|h_{\mu\nu}| << 1$ are the metric perturbations.

The metric perturbations are not unique. If the gravitational effects are small enough that one set of nearly Lorentz coordinates may be found, then many other nearly Lorentz coordinates may be found or constructed for the same spacetime. The situation is similar to the situation in electrodynamics, where the potentials associated with the fields are not unique: a *gauge transformation* may be used to convert one set of potentials to another.

In the Lorenz gauge (analogous to that used in EM theory), the Einstein equation then simplifies to

$$-\frac{1}{c^2}\frac{\partial^2 h_{\mu\nu}}{\partial t^2} + \nabla^2 h_{\mu\nu} = -\frac{16\pi G}{c^4}T_{\mu\nu},\tag{12.14}$$

that is each component $h_{\mu\nu}$ of the metric perturbation satisfies the wave equation with a source.

Far from the sources the above equations become ordinary wave equations. As the simplest example of a GW the metric perturbations we can take use

$$h_{\mu\nu} = \begin{bmatrix} 0 & 0 & 0 & 0 \\ 0 & f(ct-z) & 0 & 0 \\ 0 & 0 & -f(ct-z) & 0 \\ 0 & 0 & 0 & 0 \end{bmatrix}.\tag{12.15}$$

As is appropriate for a propagating wave, $f(ct-z)$ is an arbitrary function of $ct-z$, provided $|f(t-z)| << 1|$. The ripple in spacetime curvature has the shape determined by the function f. It is propagating with the velocity c in the z-direction without altering its shape. The spacetime is described by

$$ds^2 = -dt^2 + [1 + f(ct-z)]dx^2 + \left[1 - f(ct-z)dy^2 + dz^2\right].\tag{12.16}$$

This example satisfies linearized Einstein equations. Strictly speaking, this does not, in fact, exactly satisfy the full non-linear equations of GR. However the approximation is excellent when the amplitudes of the waves are small, which is true in all cases except near the sources. The amplitudes of the strongest GWs detectable on Earth are of order 10^{-21}.

In the presence of the sources the linearized Einstein equations are solved using the same techniques that have been discussed in the previous chapters.

PART III

Sources of Gravitational Waves

CHAPTER 13

Overview

Much has been said on the similarity between the electromagnetic and GWs in the preceding chapters. There are, on the other hand, drastic differences between the two in both the manner in which they are generated, and in the manner in which they have to be detected. The major difference is made obvious by the fact that, following the theoretical prediction, it has taken exactly one hundred years for the human species to detect a GW directly. To detect an EM wave a human needs to do no more than open her eyes.

One could say that the reason for this difference is the strength of the corresponding waves. In some sense the electric force is "stronger" than gravity by many orders of magnitude, so it should be no surprise that GWs are harder to detect.

However, even if an interaction is very weak, it is possible to produce a noticeable effect if the source is strong enough. The GWs that *have* been detected to date originated from the most energetic events in the universe. The amount of energy liberated in the first ever detected black hole merger is equivalent to 3% of the mass of the Sun. To put this into some perspective consider: the mass of all the objects in the Solar system besides the Sun itself—of all the planets and their satellites, of the comets and asteroids and the dust and the gas—together make up less than 0.2% (2 *tenths* of a percent) of the mass of the Sun.

This enormity of the energies required for generation of detectable GWs gives a hint to the usefulness of their measurements for the understanding of nature by humans. We do not have any access to such levels of energies. Occasional observations of the violent events that happened billions years ago in galaxies megaparsecs away are the only means by which we can explore such phenomena. A waveform of a GW carries the information about the properties of the objects that produced it.

Another reason for the difficulty of detecting a physical phenomenon may be its rarity.

As discussed earlier, the GWs are produced by a time-varying quadrupole moment. Physically this means a non-spherically symmetric motion of objects. A perfectly symmetric supernova explosion, for instance, would not produce a GW (however the asymmetries in these events would). In principle, GWs are emitted by any set of masses as long as they are not spherically symmetric. So, any pair of orbiting astronomical objects, such as a planet orbiting a star, produce GWs. However the strength of the emitted radiation is laughably small. Binary star formations are also common in the universe, but the magnitude of GWs produced by binary stars is also far below the level that we can ever have hope of detecting. The GWs so far detected have originated from the orbital motion of *compact binaries*: pairs of very massive and small astronomical objects, such as neutron stars and black holes.

It is also not enough to simply have compact objects orbiting one another. Potential detectability requires a binary system of two compact objects in an extremely tight orbit. For a direct detection with current tools the binary system of two compact objects has to come to an end in a *compact binary coalescence*. The pair of orbiting compact objects, initially in a large radius orbit around their common center of mass, have to slowly loose enough of their initial rotational energies to approach each other. This approach happens over the centuries as the to objects emit the energy through GWs far too small for direct detection. Over this time, the compact objects have to preserve significant amount of their mass without losing it through such common processes as one orbiting object "feeding" on the material lost by the other, and without undergoing catastrophic deformations that could "rupture" one of the objects.

The frequency of the occurrence of the detectable events depends on the evolution of galaxies. By the comparison of the predicted and actually observed detection rate we refine our understanding of this evolution.

CHAPTER 14

Black Holes

As is widely known, black holes are astrophysical objects so dense that the light cannot escape them. The idea that such objects might exist predates the era of modern physics. Newtonian physics may be used to predict the existence of "dark stars," starts that possess gravitational field so strong, that the light they radiate is unable to escape.

The wave nature of light was not firmly established until the early 19th century. The idea that light was made of particle of zero mass had been favored by Newton himself, as well as by many prominent scientists of eighteenth century. Only in year 1818 did Augustin-Jean Fresnel submitted a treatise on wave theory of light to the French academy, and Simeon Poisson and Dominique Arago, in an attempt to dispel this new theory, ended up proving it experimentally.

Suppose, as Newton did, that the light is made of massless classical particles. The acceleration due to gravity is independent of the mass of the accelerating object. It stands to reason that even the particles of zero mass should experience acceleration. John Michell and Henry Cavendish around 1783, Pierre LaPlace in 1796, and von Soldner in 1801 all carried out the estimates for the bending of light by a gravitating mass. Using the same principles as in the study of scattering in classical mechanics, they deduced that a beam of light passing a distance R from the center of an object of mass M will be deflected by an angle

$$\Delta\theta_{Newt} \approx \frac{2GM}{Rc^2}.$$

GR gives the result that is exactly twice as large:

$$\Delta\theta_{Newt} \approx \frac{4GM}{Rc^2}.$$

This latter result agrees with the experimental measurements, and was seen as one of the crucial experimental confirmation of GR in 1919.

Newtonian mechanics-based calculation of the properties of "dark star," on the other hand, leads to certain correct predictions, even though their meaning is completely altered in GR. The Newtonian escape velocity V_e is obtained from conservation of energy,

$$\frac{GMm}{r} = \frac{mv_e^2}{2},$$

so a particle of any mass m (even if $m = 0$) cannot escape to infinity if

$$r < \frac{2GM}{v^2}. \tag{14.1}$$

The light, whose "initial" velocity is c, cannot escape if it iss emitted at a distance $r \leq R_S$ where

$$R_S \equiv \frac{2GM}{c^2}. \tag{14.2}$$

So a star of mass M whose radius is less than $R_S = 2GM/c^2$ would be dark.

The radius R_S is called the *Schwarzschild radius* or *gravitational radius*. In GR this quantity takes on a more profound attribute. The only possible spherically symmetric solution of the Einstein equation, called the *Schwarzschild metric*, possesses a singularity at R_S. It is the radius of *event horizon* of the black hole. It would be possible, in principle, to obtain information from a Newtonian dark star even if we demanded that nothing can be launched with the initial speed larger than c. The light emitted from below the gravitational radius of the Newtonian dark star would not escape to infinity, but it would cross the gravitational radius. An outside observer would then be able absorb the light rather than let it "fall" back down, thus obtaining the information from below.

Nothing of the sort is possible in the case of real black holes. Light emitted inside the Schwarzschild radius starts "falling" at the instant of emission.

The modern understanding of black holes did not emerge even immediately after 1916, when Karl Schwarzschild introduced the Schwarzschild metric as the first explicit solution to the just discovered GR. Einstein himself did not believe that objects that are smaller than their gravitational radius could exist, neither did other scientists. The theoretical description of *gravitational collapse* that would produce a black hole was provided in 1939 by J. Robert Oppenheimer and his student Hartland Snyder [23], but the name "black hole" was coined only in 1967 by John Wheeler. Until then these objects were referred to simply as "collapsed stars," and were considered to be, simultaneously, exotic, speculative, and boring. The 1960s saw the discoveries of many astrophysical objects of immense power. The first *quasar*, that is the "quasistellar object" whose luminositiy exceeds the luminosity of an entire ordinary galaxiy, was identified in 1963. That discovery led to the realization of the fact that very massive objects can actually *liberate* enormous quantities of energy. The matter located outside the Schwarzschild radius undergoes massive shocks. The energetics of the strong gravitational fields associated with black holes is such that they transform up to a half of all the in-falling matter into radiation. The only more efficient process of energy generation in nature is the process of annihilation of matter and anti-matter.

Presently, black holes are detected through the observation of phenomena associated with the in-falling matter, such as accretion disks and emerging jets.

14.1 TYPES OF BLACK HOLES

14.1.1 SOLAR MASS BLACK HOLES

It is now known that there are several distinct populations of black holes. Oppenheimer and Snider predicted the existence of *stellar mass black holes*. If the mass of the remnant of a dying ordinary star exceeds *Tolman–Oppenheimer–Volkoff limit*, the star will collapse into a black hole. The masses of these black holes are between 3 and 20 solar masses (6 to $40 \times 10^3 0$ kg).

14.1.2   SUPERMASSIVE BLACK HOLES

The other type of black holes is the *supermassive black holes*, which clock in at $10^6 - 10^9$ solar masses. It is now believed that these objects exist at the centers of most galaxies, and are responsible for the luminosities of quasars.

14.1.3   PRIMORDIAL BLACK HOLES

Finally, there is a theoretical possibility of *primordial black holes*. These are very small (less than 10^{12} kg, the mass of a comet or a terrestrial mountain) objects that may have been formed at very early stages of the evolution of the universe.

CHAPTER 15

Neutron Stars

Neutron starts are smallest and densest stars known. Their radii may be estimated theoretically to be around 10–12 km (the radius of an average human city), while their typical masses are 1.4 to 2 times the Solar mass (note that the mass of the Sun accounts for 99.86% of the mass of the entire Solar System).

Obviously, an optical observation of such a star in isolation is a very difficult. Their theoretical existence was proposed in 1933 by W. Baade and F. Zwicky [21], they were discovered as *pulsars*: emitters of periodic pulses of radio waves in 1968 [22], and the first optical observation of an isolated neutron star occurred only in 1997 by Hubble Space Telescope. As with other compact objects, the primary mode of observation of these stars is not optical.

Neutron stars consist of *netron star matter*: matter so highly compressed by the gravity that its constituent elementary particles reside in the degenerate state. The matter in degenerate state cannot generate energy, therefore a neutron star can only radiate its remnant heat over millions of yours. They are hot, with temperatures around 10^6 K, but due to the small size still posses a very low luminosity. (The other type of degenerate stellar remains, the *white dwarfs*, are similarly incapable of generating energy. However, white dwarfs had been known since year 1783 simply because their volume is comparable to the volume of the Earth, which makes them much more visible.)

The best-studied neutron stars are the pulsars: the sources of persistent periodic radio signal. The reason for these radio signals lies in the fact they rapidly rotate, as fast as 1000 times per second. It is a well-known fact that the Earth's *magnetic* poles are not in the same location as Earth's geographic poles (that is, locations where the Earth's rotational axis intersects the surface). So, Earth's North magnetic pole is approximately 1,600 km south of the geographic North pole. Similarly, the magnetic axis of a neutron star is not necessarily aligned with its rotational axis. A rapidly rotating neutron star emits a beam of radio waves from its magnetic poles. When the rotational and magnetic axes of the neutron star are not aligned, the beams sweep a conical surface through space. If there is a detector anywhere on that conical surface, the beam of radiation then sweeps across the detector, producing the *lighthouse effect*: a flash of radiation detected at very regular intervals. Depending on the rotational frequency of the emitter neutron star, the observed periods are different, ranging from milliseconds to seconds. The important fact is that the period of each individual pulsar is extremely stable, with the regularity of the observed radio flashes rivals that of the atomic clocks. Some of the periods (such as that of the Hulse–Taylor) are known to the precision of 14 significant figures.

More precisely, the periods very slowly increase as the stars lose their rotational energy due to a variety of processes. However, the rate of the loss is also extremely stable, and the period of pulsar flashes is extremely predictable.

There are at least four different ways in which neutron stars enter GW research. Two of them are associated with *neutron star binaries*, a pair of neutron stars orbiting each other. Such a binary is an emitter of GW.

1. The first indirect observational confirmation of the existence of GWs was provided by a neutron star binary [24]. The orbital period for that binary was found to decrease with time. This indicates that the two stars are slowly losing kinetic energy. The rate of this loss of energy exactly agreed with the rate of loss of energy due to the gravitational radiation.

2. Much more recently, a pair of orbiting neutron stars has also provided the first joint gravitational and electromagnetic observation of an astronomical event. GW detectors heard the GW emitted from the coalescing binary NS at the same time as conventional telescopes observed the same event in the electromagnetic spectrum.

3. If a neutron star produces an asymmetrical supernova, the GW may also be generated. This possibility is somewhat speculative, but there is both observational and numerical evidence for such non-spherical supernovae.

4. Finally, rather than being the sources of GWs, neutron star pulsars may be used as a type of GW detector. This possibility relies on the immense regularity of the pulsar's timing. Tiny changes in the time of arrival of the pulses to Earth may be measured, and the cause of these changes may then be investigated. A GW propagating between the Earth and the pulsar alters the spacetime metric, and, therefore, acts as one such cause.

Unlike black holes, neutron stars are made of matter, have internal structure, and are capable of being affected and deformed by other astronomical objects. In particular, the individual stars within neutron star binaries deform each other. Such deformations have to be taken into account when considering the GW radiation.

CHAPTER 16

Early Universe

The most widely accepted theory of the origin of the universe is the Big Bang. It describes the expansion of the universe from the state of singularity into the present-day universe.

In 1917, Einstein tried to apply his recently discovered equations to the universe as a whole. He found that the original theory does not describe the static universe. He considered this problematic, and introduced a modification to the equations which would ensure that static universes are possible.

In 1922, A. Friedman proved that an expanding isotropic universe is mathematically consistent. These solutions were later re-derived and developed by G. Lemaître, H. P. Robertson, and A. G. Walker, and now bear their names. G. Lemaître noted in 1927 that an expanding universe could be traced to a single event in time.

This theoretical work became relevant in 1929, when Hubble discovered that the galaxies, in fact, move apart: the universe does, in fact expands.

The 1964 discovery of the Cosmic Background Radiation (CBR) by A. Penzias and R. Wilson provided a very strong evidence that the universe originated in a very hot dense state. This fact was conclusively proven in the early 2000s when the measurements of the polarization of the CBR left no room for the steady state theories. The measurements of the properties of the CBR are currently one of the most active areas of research in cosmology.

The Big Bang theory continued to present several theoretical problems. Three such problems: *the horizon problem*, *the magnetic monopole promlem*, and the *flatness problem* found an attractive resolution in *inflationary theory*.

The horizon problem arises from the difficulty in explaining the apparent uniformity of the universe. More specifically, the CBR arriving from different directions could be expected to have non-uniform temperature if the standard Big Bang theory were correct. However, the temperature of the CBR is very uniform.

The standard Big Bang theory also arrives at the result that the universe should contain many magnetic monopoles—in fact, the matter should be dominated by such objects. Contrary to this expectation, no magnetic monopole has been discovered ever, in spite of intense search over several decades.

Finally, it is surprising that the spacetime curvature of the universe as a whole is very close to zero. There are no reasons for the universe to be flat according to the standard Big Bang theory. The reality that it is, in fact, flat, is the flatness problem.

To resolve these theoretical problems A. Guth, A. Starobinsky, A. Linde independently introduced the theory of *cosmological inflation*. This theory postulates that shortly after the Big Bang, the universe underwent a period of rapid exponential expansion. This period lasted for an extremely short by human from 10^{-36} s after the initial singularity until approximately 10^{-32} s after the singularity. During this extremely brief period the linear dimensions of the universe expanded by a factor of 10^{26}.

Inflationary theory is currently widely accepted among the cosmologists because it resolves the problems present in the classical Big Bang theory. The theory also makes several predictions, among which is the production of background of large-scale stochastic GWs during the expansion.

The scale of the expansion postulated in the theory of inflation sets the scale for the wavelengths of the GW that should have been generated: around 10^{25} m, which corresponds to a billion light years. Correspondingly, the periods of oscillation of such GW measure in billions of years.

PART IV

Detection of Gravitational Waves

CHAPTER 17

Overview

A number of instruments have been built to detect GWs, and an even large number have been proposed. A large number of top scientists spent their careers coming up with the ideas for these detectors, building them, studying their properties, upgrading and modifying them. Only one type of these detectors has been able to achieve direct observation, about 40 years after the idea for the detector was conceived, and 18 years after the groundwork for building that detector was initiated. Several other types of detectors have either been abandoned. Several other types have now spent over a decade in preliminary development.

The designers of any GW detector have to solve one main problem familiar in other areas of technology: detecting a very weak signal in the sea of noise. The questions that have to be answered when solving this problem are, always, two-fold: is it possible to

1. amplify the weak signal so as make it measurable, while at the same time

2. *not* amplify the noise so much that the weak signal is completely buried in it.

   The answers to these questions require thorough understanding of

1. the expected signal, its strength, waveform, spectrum and evolution

2. the detector required to register the signal, and its response to any input

3. the noises which affect the detector, their features

CHAPTER 18

What Should We Measure?

The first question that needs an answer, even before we can discuss the potential signals, is: what physical quantity should we be measuring? A GW is the wave of changing spacetime metric, but can we measure the spacetime metric directly? Or, is it better to try to measure some other quantity which is affected by the spacetime metric?

It is fair to say that in most physical measurements we rely on the measurement of some quantity other than the actual quantity we want to find. To take the most trivial and basic example: to measure the magnitude of some force we would likely be resorting to measuring some sort of length, perhaps the magnitude of a stretch of a calibrated string. Similarly, to measure the amplitude of a sound wave we often resort to the measurement of the displacement of some membrane (such as the one in our ears). A measurement of the magnitude of an EM wave heavily depends on the frequency and the amplitude of the EM wave in question, but usually comes down to the measurement of electric current, or of temperature.

Underlying all such actual measurements, however, is the understanding of the effect that the actual quantity of interest can produce. An acoustic wave is a pressure wave, and pressure gradients produce longitudinal forces which can affect the positioning of thin mechanical membranes—therefore the measurement of the position of a membrane is useful for measuring sounds. An EM wave is a wave of changing electric fields, which can excite charged particles—therefore the measurements of quantities affected by the velocities of charged particles in matter leads to a reliable measurement of an EM wave.

Being the wave of distortions of spacetime metric, GW directly affect the spacing between masses in free fall. This is their most direct natural effect, and so the measurement of the spacing between two free-falling *test masses* (which we abbreviate "TM" from now on) would be the most direct natural measurement of the GWs.

There are several difficulties with such a proposition. For one, it is obviously difficult to come up with free-falling TMs whose separation may be measured for an extended period of time on the surface of the Earth. Besides that, even if we, somehow, manage to perform such a measurement, the separation between such two masses is likely to be affected by other factors much more dramatically than by extremely weak GWs.

Another way of thinking about GWs is by considering the forces they may impart on objects. As we have discussed before, the correct way of thinking about gravitational fields is through the concept of tidal forces. A GW produces tidal forces in any solid object through which it passes. A possible measurement then would be the measurement of the stresses induced in solid objects by these tidal forces.

The difficulty in such a scheme, is, again, the tiny magnitude of the tidal forces that are set up.

These two effects—the change in the separation of objects in free fall, and the tidal forces, are the only physical manifestations available for direct detection. There are some indirect measurements that can be performed, most notable of which is the fact that the energy is carried away by the GWs.

The first observational evidence of the existence of GWs came from such an indirect measurement of the energy being lost by a pair of astronomical objects to GWs, as is discussed in Chapter 20. The attempts at the detection of tidal forces set up by passing GWs in solid objects had been carried out for several decades at the end of the 20th century, as discussed in Chapter 22. All other measurements schemes rely on the measurement of the separation between two points in space. The first actual direct detection came from the observation of separation between suspended TMs. The measurements of the separation between untethered TMs in a free fall in space is being currently developed—the masses will be free falling onto our Sun in the same sense that the Earth and other planets are free falling onto our Sun as they continuously orbit it. The ideas behind this measurement are described in Chapter 25. The same quantity, though measured in a different way, is described in Chapter 26. Finally, another concept, described in Chapter 27 calls again, the measurement of separation between points in space using, in effect, radio waves "in free fall."

CHAPTER 19

The Signals

An idealized wave motion is represented by a continuous sinusoid, of constant frequency, without a beginning or an end. The only quantities required to describe it are the frequency, the amplitude, and, perhaps, the phase angle at some chosen time.

Any real wave motion, be it gravitational, electromagnetic, or any other kind, has other attributes. At some point in time the wave has to be begin, even if that time is very far in the past. At some point it has to end. During its inception the wave had to grow in amplitude over a certain time duration. The reverse has to be true at the end. Additionally, the frequency of the wave may change during its time of existence, over some additional time intervals.

GWs are commonly classified in two different ways based on the above characteristics. On one hand, different astronomical events produce GWs at different frequencies. Different instruments (past, present, and proposed) are designed to be sensitive in particular frequency bands, so the frequency classification is, at the same time, the classification of detectors (Table 19.1).

On the other hand, the produced waves never have constant unchanging wavelength. The GWs therefore are also broadly categorized into short-lived/long-lived and well-defined/poorly known categories. Different astronomical events produce GWs of different duration, and different detector designs are naturally more or less suited either to the detection of long- or of short duration.

19.1 INSPIRALS

Emitted GWs take away the energy and angular momentum of an orbiting binary system and change the frequency of orbital rotation ω. Let us recall the laws governing the classical behavior of two objects whose masses are m_1 and m_2. Let these two objects attract each other with force $k S_1 S_2 / r_s^2$, where r_s is the separation between these massive sources.

Traditionally such systems are called Keplerian, and their behavior is described with generalized Kepler's laws:

1. The two objects move in elliptical orbits, with one of the foci of the orbits located at the center of mass of the system.

2. The angular momentum of the system is constant.

3. The square of orbital period for the system is proportional to the cube of the semi-major axis:

$$r_s^3 \propto T^2. \tag{19.1}$$

Table 19.1: GWs and GW detectors at various frequencies

10^{-18} Hz – 10^{-15} Hz	Stochastic background	Studies of cosmic microwave background
10^{-9} Hz – 10^3 Hz	Supermassive ($M \propto 10^9 M_\odot$) BH binaries	Pulsar timing arrays
10^{-3} Hz – 1 Hz	Supermassive ($M \propto 10^3 M_\odot$ – $10^9 M_\odot$) BH binaries	Space-based detectors (LISA)
1 Hz – 10^3 Hz	BH and NS binaries ($1 M_\odot$ – $10^3 M_\odot$), Supernovae, Pulsars	Terrestrial interferometers, resonant mass (bar) detectors

Most of the internal properties of such system may be expressed through just two parameters. In particular, given the masses and source strength, the energy of the Keplerian system may be expressed just through the rotational frequency.

To find out what the effect of radiative loss of power on such a system we use *perturbative approach*. This means that we write down the total energy as if the system were completely stable, then consider how the properties of the system will change if this energy is slowly altered.

To build the intuition, we will approach this process from several levels of complicatedness. Let us first consider the functional dependence of changing energy, separation, and frequency.

19.2   FUNCTIONAL DEPENDENCE

In the 0th approximation the total energy of the system is constant. It consists of the size-dependent the potential energy $U = k S_1 S_2 / r_s$, and of the the frequency-dependent kinetic energy, $\Sigma m \omega^2 r$.

We can use Kepler's third law to relate the size of the binary system to the period. In terms of frequency:

$$\omega^2 \propto r_s^{-3}. \tag{19.2}$$

Without worrying about exact coefficients, we can use this proportionality to express the total energy in terms of size of the system only:

$$E \propto \frac{1}{r_s} \tag{19.3}$$

(equivalently, we could have instead expressed the energy in terms of only frequency).

Turning now to the energy loss due to radiation, we recall that the luminosity of nth multipole has the functional dependence on both the size and the frequency of the source:

$$P \propto r_s^{2n} \omega^{2(n+1)} \tag{19.4}$$

(see Section 6.5).

Using Kepler's third law proportionality Eq. (19.2) again, we can rewrite the luminosity just in terms of r_s (or just in terms of ω). So, in terms of the size,

$$P \propto \frac{1}{r_s^{n+3}}. \tag{19.5}$$

The luminosity describes how much energy is lost from the system over time: $dE/dt = -P$. The benefit of expressing both the total energy and the luminosity in terms of just one variable is that now we can write and solve an equation for just that variable. $dE/dt = -P$ then becomes

$$\frac{d}{dt}\frac{1}{r} \propto \frac{1}{r_s^{n+3}}. \tag{19.6}$$

This is a simple differential equation, whose solution may be written as

$$r_s \propto (t_f - t)^{1/(n+2)}. \tag{19.7}$$

Written in this form the solution makes it easy to interpret the result: the size of the system shrinks with time until the time t_f.

We can now use Eq. (19.2) again to find how the frequency of rotation changes with time:

$$\omega \propto (t_f - t)^{-3/2(n+2)}, \tag{19.8}$$

and we can use Eq. (19.5) again to find how the luminosity changes with time:

$$P \propto (t_f - t)^{-\frac{n+3}{n+2}}. \tag{19.9}$$

For quadrupole ($n = 2$) this means that the distance between the objects decreases as the 4th root of time, while both the frequency and luminosity increase—that is the origin of the "chirp":

$$r_s \propto (t_f - t)^{1/4}, \tag{19.10}$$

$$\omega \propto r_s^{-3/2} \propto (t_f - t)^{-3/8}, \tag{19.11}$$

and

$$P \propto r_s^{2n}\omega^{2(n+1)} \propto (t_f - t)^{-5/4}. \tag{19.12}$$

As the time approaches the "coalescence time" t_f and the separation between the sources decreases, both the radiated power and the roational frequency increase.

19.3   DIMENSIONAL ANALYSIS

The above analysis provides the time dependence the signal. We can now use dimensional analysis to evaluate how the signal depends on the nature of the sources.

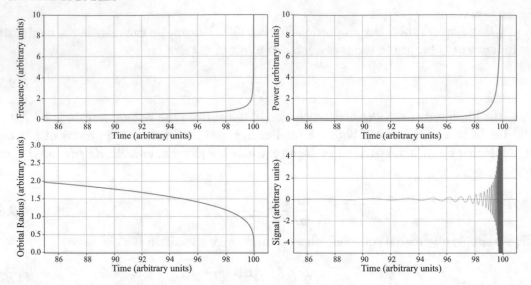

Figure 19.1: Functional forms of the parameters of an inspiral: frequency, orbital radius, radiated power, radiative signal.

19.3.1   THE MOTION WHEN RADIATIVE LOSSES ARE NEGLECTED

Let's first find the relationship between the frequency of rotation and the typical distance between the sources. We know that the answer is given by Kepler's third law, Eq. (19.2). The dimensional analysis can provide some information in addition to the functional dependence.

We should expect that the frequency, energy, and angular momentum of the rotating binary system depend on

- the force constant k,

- the two source strengths S_1 and S_2,

- the two separate inertial masses m_1 and m_2, and

- the typical distance between them r_s.

Since the source strengths only produce the forces and interaction energies, they can only ever enter any possible calculations as a product $S_1 S_2$.

The situation with the masses is more complicated. Our system of two masses is assumed to be isolated, so the center of mass of the system is stationary. This means that the momenta of the two masses are equal in magnitude. Then the velocity of the smaller mass has to be larger. If we try to calculate kinematic quantities, such as period or frequency of rotation we should find that we need to use considerations beyond dimensional analysis.

The simplest option then is to postulate some effective mass for the system, M_e, and use that until we perform more sophisticated calculations.

For the rotational period T we now write

$$[T] = [k]^\alpha [S_1 S_2]^\beta [M_e]^\gamma [r_s]^\delta. \tag{19.13}$$

Evaluating the values of α, β, γ and δ (by the same process as was illustrated in Chapter 6.5) we obtain

$$r_s^3 \propto \frac{k S_1 S_2}{M_e} T^2. \tag{19.14}$$

We obtain a version of Kepler's third law of orbital motion: the square of the orbital period is directly proportional to the cube of the typical size of the system. In terms of frequency this equation is

$$\omega^2 \propto \frac{k S_1 S_2}{M_e} r_s^{-3}. \tag{19.15}$$

We see an elaboration of Kepler's third law: the dependence of the frequency on the force of mutual attraction between the sources, and on some function of masses is can now bee seen.

19.3.2   THE ORBITAL CHANGES DUE TO RADIATION

Assuming the functional dependence of the orbital parameters on time as calculated in the previous section (Eq. (19.7)), let us now find what additional information dimensional analysis can provide.

In addition to time, r_s should also depend on the force constant k, product of source strengths $S_1 S_2$, effective mass M_e, and, since radiative losses are involved, on the speed of light c.

Bringing these together with the time dependence, we form

$$[r_s] = [t]^{1/(n+2)} k^\alpha ([S_1][S_2])^\beta [M_e]^\gamma [c]^\delta.$$

Evaluating α, β, γ and δ, we get

$$r_s \propto \left[\left(\frac{k S_1 S_2}{M_e c^2} \right)^{n+1} c(t_f - t) \right]^{1/(n+2)}. \tag{19.16}$$

For the quadrupolar radiation ($n = 2$), and the system bound by gravitational attraction ($k S_1 S_2 = G m_1 m_2$) these equations become

$$r_s \propto \left[\frac{G^3}{c^5} \left(\frac{m_1 m_2}{M_e} \right)^3 (t_f - t) \right]^{1/4}. \tag{19.17}$$

19.3.3   FREQUENCY AND CHIRP MASS

Let us now find the frequency associated with the motion and radiation. The frequency of radiation is twice the frequency of rotation, but numerical factors are not derivable from dimensional analysis anyways.

Using Kepler's third law Eq. (19.15), we obtain

$$\omega \propto \left[\frac{c^3}{(k^2 G^3)^{1/5}} \frac{1}{(S_1^2 S_2^2 M_e)^{1/5}} \right]^{5/8} \frac{1}{(t_f - t)^{3/8}}. \tag{19.18}$$

The masses and source strength go in this expression in a particular combination: $S_1^2 S_2^2 M_e$. When the system is bound by gravitational attraction ($k S_1 S_2 = G m_1 m_2$) this combination becomes $m_1^2 m_2^2 M_e$.

The *chirp mass* of a binary which produces GW radiation is defined as

$$M_{\text{chirp}} \equiv (m_1^2 m_2^2 \mu)^{1/5}, \tag{19.19}$$

where μ is the two-body reduced mass of Eq. (5.17). The above derivation shows the origins of this expression. In terms of chirp mass the frequency associated with gravitationally bound binary system is proportional to

$$\omega \propto \left[\frac{c^3}{G M_{\text{chirp}}} \right]^{5/8} \frac{1}{(t_f - t)^{3/8}}. \tag{19.20}$$

19.4   DIRECT DERIVATIONS FOR BINARIES

19.4.1   CIRCULAR MOTION, IDENTICAL MASSES

Let a binary made up two identical masses m in a circular orbit of radius R around the point half-way between them. The distance between the masses $r_s = 2R$, and the total mass of the system is $M = 2m$.

The two masses keep circling their common center of mass because the force of attraction between them acts as the centripetal force

$$k \frac{S_1 S_2}{(2R)^2} = m \omega^2 R$$

which leads to a form of Kepler's third law:

$$\omega^2 = k \frac{S_1 S_2}{4m} R^{-3}. \tag{19.21}$$

The total energy of this system is

$$E = 2 \frac{m v^2}{2} - k \frac{S_1 S_2}{2R} = m \omega^2 r^2 - k \frac{S_1 S_2}{2R},$$

and with Eq. (19.21) this simplifies to

$$E = -k\frac{S_1 S_2}{4R}.$$ (19.22)

This is an example of the *virial theorem* of classical mechanics:

$$E = \frac{PE}{2} = -KE$$

To express the power loss by the system we use its multipole moments. The multipole moments of a binary and their derivatives have been calculated in Section 6.4.1. The expression for luminosity

$$P = \frac{128}{5}\frac{Gm^2}{c^5}R^4\omega^6$$

can now be combined with Kepler's third law to produce

$$P = \frac{2}{5}\frac{G}{c^5}\frac{(kS_1 S_2)^3}{m}R^{-5}.$$ (19.23)

We can now set the energy loss condition $dE/dt = -P$:

$$\frac{d}{dt}(-k\frac{S_1 S_2}{4R}) = -\frac{2}{5}\frac{G}{c^5}\frac{(kS_1 S_2)^3}{m}R^{-5}$$

or

$$R^3\frac{dR}{dt} = -\frac{8}{5}\frac{G}{c^5}\frac{(kS_1 S_2)^2}{m}.$$ (19.24)

Solving this we get

$$R(t) = \left[\frac{32}{5}\frac{Gk^2}{c^5}\frac{S_1^2 S_2^2}{m}(t_f - t)\right]^{1/4}.$$ (19.25)

Using Kepler's third law again we obtain the frequency of rotation:

$$\omega = \frac{1}{2}[2\cdot 5^3]^{1/8}\left[\frac{c^3}{(k^2 G^3)^{1/5}}\frac{1}{(S_1^2 S_2^2 m)^{1/5}}\right]^{5/8}\frac{1}{(t_f - t)^{3/8}}.$$ (19.26)

The frequency of GW radiation is twice the oscillation frequency.

19.4.2 CIRCULAR MOTION, UNEQUAL MASSES

The above analysis goes without major changes to the case where the two masses are not equal. Classical mechanics predicts that two unequal masses do not maintain circular orbits: instead they travel along elliptical paths. However, it can be shown that the constant energy loss always "rounds out" the orbits.

The only alteration to the formulas derived above is the replacement of the mass m by the two-body reduced mass μ everywhere in the analysis. The results are

$$f_{GW} = \frac{5^{3/8}}{8\pi} \left[\frac{c^3}{GM_{\text{chirp}}} \right]^{5/8} \frac{1}{(t_f - t)^{3/8}}, \tag{19.27}$$

and

$$h_0 = \frac{1}{r} \left[\frac{5G^5 M_{\text{chirp}}^5}{c^{11}} \right]^{1/4} \frac{1}{(t_f - t)^{1/4}}, \tag{19.28}$$

where M_{chirp} is defined in Eq. (19.19).

19.5   FULL INSPIRAL WAVEFORM

The above discussion uses Newtonian description of gravity and motion everywhere, except where the slow energy loss is postulated. Such adiabatic description can possibly work only in the non-relativistic regime, when the velocities of the orbiting bodies are far below relativistic speeds and when the gravitational interaction energies are far below rest mass energies. In the later stages of the inspiral such approximations are inadequate.

The time evolution of a binary system is usually split into three stages. The first is the *adiabatic inspiral* stage described above. Following the adiabatic inspiral stage is the *merger* or *plunge*. During this stage the two compact objects plunge together to form a single black hole. In this stage it is necessary to use Einstein Equations for computations.

Such computations are extremely difficult. Two approximation methods are used. One is numerical relativity, that is the numerical solution of the Einstein equations. The other method employs increasingly accurate post-Newtonian and post-post-Newtonian approximation schemes. Post- and post-post-Newtonian approximations extend the calculations to increasing orders of the relativistic parameter $(v/c)^2$.

The last is the *ringdown* phase, which is modeled by *quasi-normal modes* of a single black hole.

19.6   BURSTS

According to the dictionary the word "burst" means "to break open suddenly from pressure within," or "to emerge suddenly." Both definitions are meaningful in physics. The relevance of the definition which already contains another physics word (pressure) is obvious. But when a scientist talks of "a burst of radiation" the reference is more to the sudden emergence, than to a physical explosion.

Both of these definitions of the word "burst" have been used in GW research as well. One may talk about a burst of gravitational radiation, that is about a relatively brief amount of time during which the magnitude of GW is relatively large. Or one may, instead, have in mind an event, such as a supernova explosion, which would generate a GW as a result of a "burst" of a star.

The usage of the word has evolved somewhat over the years. The strength of the emitted wave has to be sufficiently large to be detectable by the human-made devices. In all direct detections to-date, the sufficiently high signal-to-noise ratio has lasted no more than a few fractions of a second. Such signals certainly should qualify as "suddenly emerging bursts of radiation" as far as their visibility goes.

On the other hand, these signals have come from the best understood sources of gravitational radiation. We know that they come from the end stages of the inspiral of a binary, which has been radiating weaker GW for a long time, before finally coming to the stage where its radiation became strong enough to be detectable by humans.

So the definition of "a GW burst" has somewhat evolved over time. Before the advent of GW interferometers it was more common to classify any short GW signal as a burst.

A classic criterion for classifying a GW signal as "burst" relied on the comparison of the duration of the event with the amount of time necessary for the observer's motion to affect the detected frequency. For the observer on Earth this means that the duration of the "burst" should last less than the amount of time necessary for the Doppler shifts due to Earth's orbital motion to become significant. Doppler effect is approximately given by $f_o = f_s(1 + v/c)$, so the changes in the observed frequency are proportional to the changes in velocity. Over the observation time τ the Earth changes its velocity by $\Delta v = a_{centripetal}\tau = \Omega_E^2 R_E \tau$. A version of the Nyquist theorem states that the finest frequency resolution possible within time τ is $2/\tau$. Therefore, the observed frequency is affected by the Earth's rotation when $2\tau > f\Omega_E^2 R_E \tau$.

The above calculation gives the observation time τ of the order of an hour. All directly observed GWs are classifiable as "bursts" under such definition. This makes such a designation sufficiently broad to be not very useful. Therefore a different definition of a burst is in current usage.

Presently, a GW burst refers to a short-lived *poorly understood* category of GW signals. This still implies that the duration of the signal is short. Additionally, however, it also implies a violent astrophysical event other than a collision and merger of two compact objects. The most obvious of such events is stellar explosion or a an implosion, a supernova—a "burst" of a star.

A spherically symmetric explosion/implosion contains only monopole source, and, therefore, would not result in a release of GWs. Fortunately, there is evidence that many supernovae explosions involve asymmetry.

The unfortunate part about the bursts is that, by the definition ("poorly understood"), the shape of the signal from such event is unknown. Instead of hunting for a particular signature in the data, one has to rely purely on the idea of coincidence: multiple detectors should all register the same source, and the time delay between the signal arrival should point to the direction of the arrival.

19.7   STOCHASTIC SIGNALS

The universe is filled with GWs. As mentioned previously, they are emitted by any set of non-spherically symmetric masses. The magnitudes of most of these GWs are miniscule and they will never be detectable. However, there is number of GWs propagating through space of sufficient amplitude above any minimum detectability limit. As new detector devices are invented and built, the detectability limit will become lower, and more of these GWs are expected to become observable.

Together, the multitude of GWs produced by different sources, propagating in different directions with different amplitudes and polarizations constitute the *stochastic GW background*.

It is useful to have a description of this background.

Since we lack the knowledge of each individual GW in this background, the description is necessarily *statistical*. The amplitude, direction, and polarization of each individual GW within the background has to be treated as a *random* variable.

Before returning to GWs, it is best to understand the situation in the simpler case: a stochastic background of scalar waves, for example, of sound waves. How would we describe the air pressure at some point within an enclosed, thermally insulated, but noisy room?

As discussed in Section 1.2, each individual wave may be described as

$$f(x,t) = \text{Real}\{Ae^{i(x \pm vt)}\} = \text{Real}\{Ae^{i(\mathbf{k}\cdot\mathbf{x} \pm \omega t)}\}, \tag{19.29}$$

where the complex amplitude A includes the description of both the amplitude and of the phase of the wave.

A multitude of waves may be obtained by summing such individual waves, producing the Fourier sum:

$$\Sigma_k (A_k e^{i(\mathbf{k}\cdot\mathbf{x} - \omega t)}). \tag{19.30}$$

When frequencies of different waves may differ an arbitrarily small amount, the Fourier sum becomes Fourier integral:

$$\Phi(t, \mathbf{x}) = \int A_k e^{i(\mathbf{k}\cdot\mathbf{x} - \omega t)} d^3\mathbf{k}. \tag{19.31}$$

Since we are describing a stochastic wave field, the complex amplitudes $A(\mathbf{k})$ are random variables. There is a certain probability that at some moment t_1 at the location $\mathbf{x}_1$ the amplitude and the phase of the wave field is given by values $\Phi_1 = |\Phi_1|e^{-i\phi_1}$. The combination of all such probabilities is described by the joint probability density

$$p_n(\Phi_1, t_1, \mathbf{x}_1; \Phi_2, t_2, \mathbf{x}_2; \ldots; \Phi_n, t_n, \mathbf{x}_n). \tag{19.32}$$

The expectation value of the wave field at a location $\mathbf{x}_1$ at time t_1 is given by the integral sum of all possible values given their probabilities:

$$\langle \Phi(t_1, \mathbf{x}_1) \rangle = \int d\Phi_1 \Phi(t_1, \mathbf{x}_1) p_1(\Phi_1, t_1, \mathbf{x}_1). \tag{19.33}$$

CHAPTER 20

Evidence from Pulsars

Just like EM waves, GWs carry energy. This energy is carried away from the radiating system. The amount of energy carried away from the source depends on the details of the structure and dynamics of the radiating system, in particular, on the energy still remaining within the system.

The first indirect observational confirmation of the existence of GWs was obtained by careful observation of a radiating system. As the GWs removed the energy from the system the orbital rotation frequency diminished. The reduction of the frequency furnished the indirect evidence of the existance of GWs.

The astronomical system whose orbital period is most amenable to being precisely measured is a *neutron star binary pulsar*. In such a system, one of the two neutron stars is a pulsar, which, as explained earlier, means that its own rotation around its own axis creates the lighthouse effect. This provides a precise metronome, from which it is possible to calculate the masses of the orbiting stars, as well as their orbital parameters.

The first discovered, and therefore most famous, such system is the *Hulse–Taylor Binary Pulsar* PSR B1913+16 [24]. PSR denotes "Pulsing source of radiation," the numbers denote the celestial coordinates (right ascension and degrees of declination), and the letter B designates the astronomical epoch to which the coordinates are referenced. This pulsar was discovered using the Arecibo radio telescope whose 305 m diameter made it the largest radio telescope from 1963 until 2016 (when *500-meter Aperture Spherical Telescope* (FAST) was completed in China). For their 1974 discovery of this pulsar, Russel Hulse and Joseph Taylor, Jr. were later awarded a Nobel Prize.

Pulse–Taylor and other pulsars exhibit a number of general-relativistic effects in their behavior: the precession of the orbits, the relativistic Doppler effects and the Shapiro time delays. The effect that makes obvious the loss of energy to gravitational radiation is the change in the orbital period.

The masses of both the pulsar NS and of its unseen companion is about $1.4 \times M_{sun}$, and the orbital period is approximately 7.74 h. The decrease in this orbital period due to the emission of GWs is predicted from Eq. (19.27) to be 10 μs per year. Timing measurements over more than three decades of observations yield that the otbital perios has been decreasing at a rate 0.997 ± 0.002 times that predicted as a result of GW emission [25].

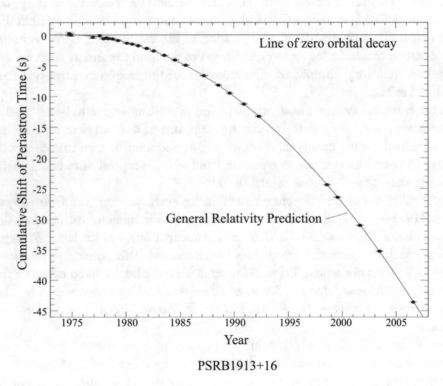

PSRB1913+16

Figure 20.1: PSRB1913+16

CHAPTER 21

Noises

21.1 OVERVIEW

By necessity, the GW detectors are built to measure some of the smallest conceivable levels of disturbance. Many effects other than the GW create disturbances of much larger magnitude. These effects range from the expected and even mundane ones to the most fundamental. Many of these may be themselves of interest to other branches of science. However, from the point of view of GW detection they represent *noise*.

On the mundane side, we encounter the disturbances due to the human activity. Walking in the proximity of a detector creates sufficient amounts of ground vibrations to make it unusable. So do unmitigated levels of cooling and heating systems. Human activity outside of the area immediately close to the detector also interferes with the detection: the detectors work best at night, when vehicular traffic on the roads surrounding the detector site is at a reduced level. Electromagnetic pickup is inevitably present at the frequencies of AC from the electrical power lines.

Such noises are inevitable wherever humans exist, namely, on the planet Earth. Any terrestrial detector is subject to these noises.

Less mundane sources of interference originate from the natural environment. Seismic vibrations are always present at any terrestrial site. The major geologic events such as earthquakes are only but one of the many sources of such vibrations. The ocean tides and the weather patterns all create such large ground vibrations that the detectors operate more optimally in certain seasons. Finally, even the quasi-static (slow) motions of distributions of matter around the detector creates additional disturbance due to a direct gravitational attraction between the test masses and those large moving masses—such disturbances create the so-called *gravity gradient noise*.

Many of these noises are naturally resolved if the detectors are based in outer space. A GW detector on the Moon, for instance, would be a subject to reduced levels of seismic noise, and the noises associated with atmosphere and oceans would not be present. Some new noise sources appear in that environment. Since the detectors are not protected by the atmosphere, they would be subject to cosmic ray and solar wind disturbances.

There are no proposals to put a GW on the Moon yet, however an artificial satellite-based GW detector is under active development. Such a detector, spread over several spacecraft is subject to the *acceleration noise*, which results from mutual motion between the spacecraft.

Finally, there are fundamental sources of non-GW disturbances. The third law of thermodynamics states that it is impossible to cool a physical system to the temperature of absolute zero. Any object—including any conceivable test mass—at a non-zero temperature is a subject to *thermal noise*. Any detection mechanism is associated with finite number of measurements: this leads to *readout noises*. Quantum mechanics guarantees that there are pairs of *conjugated variable* whose exact values are not definable at the same time, mathematically, these are variables associated with the *non-commuting operators*. In effect this means that the precise measurement of one set of conjugated variables inevitably creates disturbance in the other, leading to other fundamental sources of noise.

The job of a detector scientist is to understand every disturbance with the goal of either eliminating them or, more often, mitigating their effects by attenuating, subtracting, and compensating for their presence.

We will take a look at a few sources of noise and at the ways of dealing with them. Before that, let us understand what distinguishes the "noise" from a "signal."

21.2 DESCRIPTION OF NOISE

A potentially detectable wave motion signal can be described as a certain deterministic mathematical function of time, such as the one discussed in Section 19.1. The signal can be recognized for what it is based on this known function of time, as well as by the fact that it is detected by several detectors along its path.

A more complex situation arises when a stochastic background of signals is present, as in Section 19.7. Search for a signal in such a situation is sometimes described as "searching for noise in noise." The only way to recognize a signal in such a situation is if it is detected by several detectors along its path, as, for example, in Chapter 27.

How is the noise itself described? The noise is a random time series. There is not a single mathematical function that always describes a particular disturbance. The noise originating from the same source, under the same circumstances, can be expected to produce different disturbances.

Even though the particular disturbances are different each time, they have the same unwanted effect on the measurements. To describe a particular noise mathematically, therefore, we have to describe the mathematical properties of that noise that produce the same unwanted effects.

The unwanted effects are the spurious readings on the measuring devices. The precise time dependence of these spurious readings is different each time, however their magnitudes can be described. Because of these considerations the noise is mathematically described by it *power spectrum density*. The deviations of the measured quantity from the average are measured over a long period of time. These deviations then are effectively squared (more precisely, but beyond the scope of this book, one calculates the autocorrelation function). The magnitudes of these *squared* deviations at various frequencies is calculated. This process is repeated multiple time, and the results are averaged. The units of the obtained power spectrum density are those of the measured signal squared per frequency.

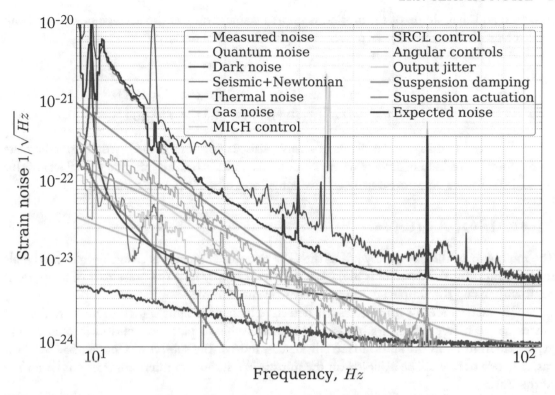

Figure 21.1: An example "noise budget," or accounting of sources of noise that combine to form the limiting sensitivity of the detectors. In this figure we show a variety of noise sources that affect the Livingston detector at low frequencies. Adopted from `https://www.ligo.org/science/Publication-O1Noise/index.php`.

It is common to take the square root of the power spectral density to obtain *amplitude spectral density* in units of the measured quantity per *square root of Hz*.

The unified picture of the the measured noises is usually presented in a *noise budget* plot, such as shown in Fig. 21.1. It is a log-log plot which shows every separate noise as a separate curve, along with the curve that show the sum of all the noise sources.

Let us look at the most obvious and at the most fundamental sources of noise.

21.3   SEISMIC NOISE

This is one of the most obvious noises present in any terrestrial detector. The ground moves everywhere on Earth, although the magnitudes of the motion varies from place to place. As described in the previous section, even at one particular point on the Earth's surface the exact direction and magnitude of the motion is unpredictable, however the average deviations from the average are described by the power or amplitude spectral density.

A typical level of noise is

$$f(x) = \begin{cases} 10^{-7} \text{cm}/\sqrt{\text{Hz}} & \text{for } 1 < f < 10 \text{ Hz} \\ 10^{-7} \text{cm}/\sqrt{\text{Hz}} \, (10 \text{ Hz}/f)^2 & \text{for } 1 < f < 10 \text{ Hz.} \end{cases}$$

The need to mitigate the effects of seismic noise is not limited to GW detectors: there are many domains of human activity where vibration is undesirable. To mitigate the effects of this noise the sensitive detectors usually involve some form of *vibration isolation*.

21.4 ISOLATION

The main part of the day-to-day job of the GW detector engineers is ensuring that the signal-to-noise ratio for the detector is high. Let us discuss briefly how the noise part of this ratio is dealt with, using the simple but immediately relevant example.

The central idea of interferometric detection is that the light is used to measure the distances between free-falling test masses. It is conceivable to place test masses into space and expect them to be in the state of free fall, that is, indeed, the subject of Chapter 25. But how can the state of free fall be achieved for the mirrors of a stationary interferometer on the surface of the Earth?

Certainly, the TMs are not rigidly attached to the ground. Instead, they are held at the ends of multi-stage pendulums, held with long fibers. The fibers themselves are not rigidly connected to the ground, but, instead, are held by a system of carefully calibrated springy metal blades. Finally, the assembly which holds the pendulums and the blades itself is not attached directly to the ground, but is mounted upon multistage seismic isolation platform.

So the TMs are not solidly attached to the ground. Still, naively it would appear preposterous to regard them as objects in a free-fall. However, in a very definite sense they are. To understand this, we first investigate what kind of isolation the multi-stage suspension systems provide from the seismic motions of the ground.

21.4.1 SIMPLE HARMONIC OSCILLATORS (SHOS)

We first review the free motion of simple harmonic oscillators. Such systems undergo simple sinusoidal motion of constant frequency ω_0. The value of the frequency depends on the parameters of the system, while the amplitude depends only on the initial conditions.

The standard examples are:

- a test mass m attached to an ideal massless spring of spring constant k. The frequency of free oscillations is given by $\omega_0 = \sqrt{\frac{k}{m}}$;

- an electric circuit consisting of an ideal capacitor C and an inductor L connected by perfectly conducting wires. In this case, $\omega_0 = \sqrt{\frac{1}{LC}}$;

- a mathematical pendulum: test mass suspended by an ideal string of length l, located in a Newtonian gravitational field of constant acceleration $\mathbf{g}$. This system can be modeled as an ideal oscillator only in the limit of small amplitude of oscillation, in which case $\omega_0 = \sqrt{\frac{g}{l}}$.

The above results follow from the classical equations of motion. So, for example, in the Newtonian description of the motion of a test mass on a spring, the only force applied to the test mass is the spring force given by the Hooke's law. Placing the origin at the equilibrium location of the mass, the elastic force is given by $F_{spring} = -kx$. Putting this force into Newton's second law results in

$$ m\frac{d^2x}{dt^2} = -kx \tag{21.1} $$

or

$$ \frac{d^2x}{dt^2} + \frac{k}{m}x = 0. $$

The solution of this equation is a sinusoid motion of frequency $\omega_0 = \sqrt{\frac{k}{m}}$. It may be written in a number of equivalent ways:

$$ x(t) = x_i \cos(\omega_0 t) + \frac{v_i}{\omega_0} \sin(\omega_0 t), $$

where x_i and v_i are the initial position and the initial velocity of the mass;

$$ x(t) = X_{\max} \cos(\omega_0 t + \phi_0), $$

where $X_{\max} = \sqrt{x_i^2 + \left(\frac{v_i}{\omega_0}\right)^2}$ is the amplitude of oscillation, and $\phi_0 = \tan^{-1}\left(\frac{v_i}{x_i \omega}\right)$ is the initial phase of the sinusoid;

$$ x(t) = C_+ e^{i\omega_0 t} + C_- e^{-i\omega_0 t}, $$

where the complex amplitude of oscillations includes both the effects of initial position and velocity; and, finally,

$$ \text{Real}\{A e^{i\omega_0 t}\}. $$

This last is the most convenient and economical way of describing oscillations. The *complex amplitude A* is a complex number whose real-valued magnitude is the amplitude of oscillations:

$|A| = X_{\max}$, and whose phase is the initial phase ϕ_0 of the sinusoid. A depends on the initial offset x_i and initial velocity v_i through

$$A = \frac{1}{2}(x_i + i v_i/\omega_0).$$

In practice, when using the last method of describing oscillations, we never write the word "Real." We write $Ae^{i\omega_0 t}$, with the understanding that after the complex quantities A and $e^{i\omega_0 t}$ have been multiplied, only the real part of the resulting expression should be taken.

21.4.2   ISOLATION PROVIDED BY SIMPLE HARMONIC OSCILLATORS

To understand the amount of isolation provided by simple harmonic oscillators, we let the support point (the "other end" of the spring or of the pendulum string) be non-stationary. Specifically, we let the support point oscillate at some frequency ω. How well the oscillator reduces the disturbances can then be understood by examining the *transfer function*. The transfer function is simply the ratio A_{out}/A_{in} of the amplitude of the motion of the test mass A_{out} to the amplitude of the motion of the suspension point A_{in}.

The only change from the setup of the previous section is that the length of the spring (or the angle of the pendulum) depends not just on the location of the mass, but additionally on the location of the suspension point. Specifically, the spring becomes longer whenever the mass moves *down* and whenever the suspension point moves *up*:

$$\text{Length of spring} = x - A_{in}e^{i\omega t}.$$

With this change we have an inhomogeneous differential equation:

$$\frac{d^2 x}{dt^2} + \frac{k}{m}\left(x - A_{in}e^{i\omega t}\right) = 0. \tag{21.2}$$

The frequency $\omega_0 = \sqrt{k/m}$ is still a useful quantity, so we should re-write the equation in terms of ω_0, rather than k:

$$\frac{d^2 x}{dt^2} + \omega_0^2 x = \omega_0^2 A_{in}e^{i\omega t}. \tag{21.3}$$

From theory of ordinary differential equations we know that the solutions of inhomogeneous equations consist of *complimentary* and *particular* solutions. Under the influence of the external disturbance, the motion described by the complimentary part of the solution dies out with time, while the motion described by the particular part persists. To find the particular solution we assume that it has the same frequency as the external disturbance, so we let $x(t) = \text{Re}\, A_{out}e^{i\omega t}$. Substituting, and taking the derivatives, we get

$$-\omega^2 A_{out}e^{i\omega t} + \omega_0^2 A_{out}e^{i\omega t} = \omega_0^2 A_{in}e^{i\omega t},$$

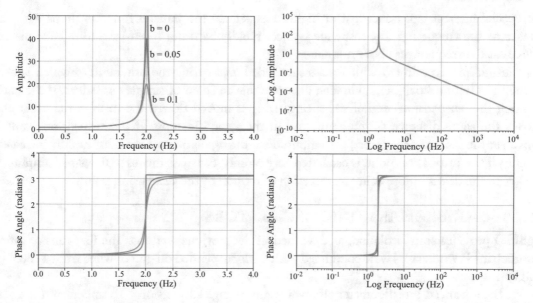

Figure 21.2: **A resonant curve.**

which simplifies to

$$\frac{A_{out}}{A_{in}} = \frac{\omega_0^2}{\omega_0^2 - \omega^2}. \tag{21.4}$$

Plotting the result, we see the well-known resonance curve.

So far our discussion of the harmonic oscillators should be very familiar from the study of mechanics. Resonance phenomena are widely studied in many branches of physics, chemistry, and engineering, and the graphs of the resonance curves are ubiquitous in the textbooks devoted to those subjects. As the graph shows, if no measures are taken to limit the motion of the test mass, its oscillations would grow at the resonant frequency without a limit.

The measures that have to be taken to limit the motion will be discussed in Section 21.4.5. Our focus in this discussion is in the frequency regions *outside* the resonance. As is obvious from the resonant curve, if the support point is vibrating at a frequency that is higher than the natural frequency of the oscillator, the amplitude of oscillation of the test mass is much smaller than the amplitude of input vibrations.

How much smaller?

Directly from Eq. (21.4), we obtain:

$$\frac{A_{out}}{A_{in}} \text{ goes to } \begin{cases} 1 & \text{if } \omega = 0 \\ \infty & \text{if } \omega = \omega_0 \\ -\omega_0^2/\omega^2 & \text{as } \omega \to \infty. \end{cases}$$

Of special interest is the fact that at higher frequencies the transfer function drops with the square of the frequency. The suspended object rapidly becomes insensitive to outside disturbances at the frequencies above resonance.

A simple experiment with a mass suspended on a spring the other end of which is held in hand, or with a small pendulum can illustrate this behavior of simple oscillators. If the hand holding the "suspension" end of the spring/string is moved slowly, the mass at the other end moves the same distance as the hand, and in the same direction. If the suspension point is moved at the resonant frequency, the amplitude of the oscillations of the mass rapidly increases. Finally, if the suspension point is oscillated very rapidly, the mass moves with a small amplitude always in the direction opposite the direction of motion of the suspension point.

21.4.3   DOUBLE HARMONIC OSCILLATORS

A SHO provides some isolation, as described in the previous sections. The first runs of interferometric GW detectors were conducted with TMs suspended on guitar-wires from a "bed" of coil springs.

The advanced interferometer all involve multi-stage suspensions. To understand the reasons we take a look at double harmonic oscillators as the simplest example of a multi-stage one.

Double harmonic oscillators (DHOs) consist of two SHOs connected sequentially. The examples of such a system include:

- two test masses m_1 and m_2, connected "in series." An ideal massless spring of spring constant k_1 connects m_1 to the support, and another ideal massless spring k_2 connects m_2 to m_1;

- an electric circuit consisting of an two LC circuits in series;

- a double mathematical pendulum: test mass m_1 is suspended by an ideal string of length l_1 from the support, and another mass (m_2) is suspended from m_1 by another ideal string of length l_2. When this system is allowed to oscillate with arbitrarily large amplitude, it becomes chaotic (in fact, it is a case-book study example of chaotic systems). In the limit of small oscillations it is still well modeled as a double harmonic oscillator.

Let us concentrate on the double-spring system. The first mass m_1 has two forces acting on it, from each of the spring, while the second mass is affected only by the second spring. Paying careful attention to the directions, we write two equations of motion for each of the two masses:

$$m_1 \frac{d^2 x_1}{dt^2} = -k_1 x_1 + k_2(x_2 - x_1)$$
$$m_2 \frac{d^2 x_2}{dt^2} = -k_2(x_2 - x_1).$$

$$(21.5)$$

Since we know that for a SHO the natural frequencies are given by $\sqrt{k/m}$, we may simplify writing the above equation by defining $\omega_1 = \sqrt{k_1/m_1}$ and $\omega_2 = \sqrt{k_2/m_2}$. Another frequency, which we define as $\omega_{21} \equiv \sqrt{\frac{k_2}{m_1}}$, also makes appearance in this system.

A number of other frequencies could be obtained by combining m_1, m_2, k_1, and k_2, for example

$$\sqrt{\frac{k_1}{m_2}} \quad \text{or} \quad \sqrt{\frac{k_1 + k_2}{m_1}} \quad \text{or} \quad \sqrt{\frac{k_1 + k_2}{m_1 + m_2}}.$$

Such additional frequencies may make appearance in the solutions of various two-spring-two-masses configurations.

With these definitions the equations are

$$\frac{d^2 x_1}{dt^2} + \omega_1^2 x_1 - \omega_{21}^2 (x_2 - x_1) = 0$$
$$\frac{d^2 x_2}{dt^2} + \omega_2^2 (x_2 - x_1) = 0. \tag{21.6}$$

To solve this system we assume that both masses move with the same frequency, but different amplitude:

$$x_1(t) = X_1 e^{i\omega t} \quad \text{and} \quad x_2(t) = X_2 e^{i\omega t},$$

where both X_1 and X_2 are complex numbers of the form $X = X_{\max} e^{i\phi_0}$. Substituting these equation, we get

$$\frac{d^2 x_1}{dt^2} + \omega_1 x_1 \quad -\omega_{21}(x_2 - x_1) = 0$$
$$\frac{d^2 x_2}{dt^2} \quad\quad +\omega_2(x_2 - x_1) = 0. \tag{21.7}$$

We can solve it by any method. The amplitudes cancel out, and we get the following equation for the possible frequencies:

$$\omega^4 - (\omega_1^2 + \omega_2^2 + \omega_{21}^2)\omega^2 + (\omega_1^2 \omega_2^2) = 0.$$

The two solutions, ω_{0+} and ω_{0-} are given by

$$\omega^2 \equiv \omega_{0\pm}^2 = \frac{\omega_1^2 + \omega_2^2 + \omega_{21}^2}{2} \pm \sqrt{\left(\frac{\omega_1^2 + \omega_2^2 + \omega_{21}^2}{2}\right)^2 - \omega_1^2 \omega_2^2}. \tag{21.8}$$

21.4.4   ISOLATION PROVIDED BY DOUBLE HARMONIC OSCILLATORS

Now let us find out how much better a DHO isolates the test mass from the disturbance in the support. Only the first Eq. (21.5) needs to be modified by introducing the additional term, as was done in Eq. (21.2):

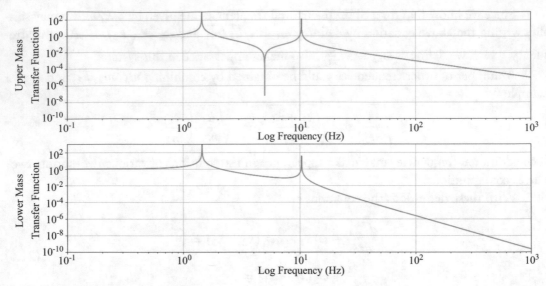

Figure 21.3: An example of the transfer functions of the two masses of a double harmonic oscillator without dumping.

$$\frac{d^2x_1}{dt^2} + \omega_1^2(x_1 - A_{in}e^{i\omega_{in}t}) \quad -\omega_{21}^2(x_2 - x_1) = 0$$
$$\frac{d^2x_2}{dt^2} \qquad\qquad\qquad +\omega_2^2(x_2 - x_1) = 0.$$

(21.9)

As in the case with SHO we want to calculate the particular solution by assuming that it has the same frequency as the external disturbance:

$$x_1(t) = A_1e^{i\omega t}$$
$$x_2(t) = A_2e^{i\omega t}.$$

(21.10)

Substituting, taking the derivatives, and working through lengthy rearrangement, we get

$$\frac{A_1}{A_{in}} = \frac{\omega_1^2(\omega_2^2 - \omega^2)}{(\omega^2 - \omega_{0+}^2)(\omega^2 - \omega_{0-}^2)}$$
$$\frac{A_2}{A_{in}} = \frac{\omega_1^2\omega_2^2}{(\omega^2 - \omega_{0+}^2)(\omega^2 - \omega_{0-}^2)}.$$

(21.11)

The plots of the two transfer functions show both resonance and anti-resonance peaks (an anti-resonance for one of the masses can happen if at a certain frequency all of excitation energy goes into the vibration of the other mass). But especially interesting to us, as with SHOs, is the behavior of transfer function outside the resonances.

Directly from Eq. (21.11) we obtain:

$$\frac{A_2}{A_{in}} \text{ goes to } \begin{cases} -1, & \text{if } \omega = 0 \\ \infty, & \text{if } \omega = \omega_0 \\ \omega_1^2 \omega_2^2 / \omega^4 & \text{as } \omega \to \infty. \end{cases}$$

The major difference from the SHO isolator is the power of the ω in the denominator of the transfer function above resonances. For SHO the transfer function decreases as ω^2, for DHO it goes down as ω^4. It is not difficult to use the same method (or simple extrapolation) to understand that each additional oscillator stage reduces the transfer function by an additional ω^2 factor. For example, the transfer function of a four-stage harmonic oscillator decreases in proportion to the *eighth* power of frequency.

This behavior is, actually, very familiar to electrical engineers. Filtering unwanted frequencies by using electronic oscillators is a major part of that profession. In the electrical engineering parlance, a SHO creates a *single pole high pass filter*, and every additional harmonic oscillator stage adds a pole.

21.4.5   DAMPING AND NOISE

The description of isolation provided by multi-stage harmonic oscillators has one serious problem: infinitely large resonances. This means that even tiny disturbances would drive test masses into ever increasing swings at the resonant frequencies. Strictly speaking, infinitely large resonances are not possible, but it is possible to have resonances large enough that the test mass would go into violently large swings.

In most ordinary situations, where the noise levels do not have to be kept extremely low, the amplitude of resonances is reduced to the finite levels by the introduction of damping. The most common form of damping is velocity-proportional damping, described by a damping coefficient $-b(dx/dt)$ in the differential equations. Such damping corresponds, for instance, to the air resistance produced when solid objects move at low speeds through most gases at normal density and pressure.

In other situations damping terms may be different. For instance, dry friction, studied in basic physics, is independent of velocity. As another example, objects moving through gases at higher speeds may experience resistance which is proportional the the square of the velocity.

The introduction of gases into the GW detector would lead to unacceptable levels of noise. In fact, providing ultra-high vacuum levels inside multi-kilometer tubes represents one of the major practical challenges and expenses in building a terrestrial interferometric detector.

Another source of damping, less obvious in the ordinary situations, is the *internal friction* of springs and wires. When solid materials is put under any kind of stress, it undergoes two types of deformations: *elastic* and *plastic*. Elastic deformations are described by various types of linear stress-strain relationships, the best known of which is the Hooke's law. Under purely elastic deformations the energy of the external applied force is transferred to the energy of separation between the atoms of crystalline structure of the solid. Once the external forces are lifted, the solid returns to its previous state.

Under plastic deformation, the energy of the external forces not only changes the separation between the atoms, but also dislocates them. Plastic deformations are irreversible, and dissipate energy. Therefore, the effects associated with plastic deformations may be described as the "internal friction."

Typically, the behavior of solid materials is well described as elastic up to certain limit called the "proportional limit." After reaching that limit the material begins to exhibit various signs of plastic deformations. So in most ordinary situations it is entirely correct to assume that Hooke's law, for example, hold precisely.

However, some amount of atomic dislocations is always present, even at the very small level of deformations. In building GW detectors the internal friction of the suspension blades and suspension fibers is carefully studied.

The upshot of the previous discussion is that there are many different potential sources of damping the unwanted infinitely large resonances. Is might appear that by choosing a proper source of damping we could achieve good filtering of the input noise at higher frequencies, while, at the same time, limit the unwanted swinging at the resonant frequencies.

This hope, however, is false. One of the major discoveries of 20th century physics is the *fluctuation-dissipation theorem*, discovered by H. B. Callen in 1951–1952. It states, in numeric terms, that any amount of dissipation in a system is associated with the random fluctuating forces. This is intuitively clear. If an energy of a swinging pendulum is dissipated by the surrounding gas, that energy is converted into the energy of random motion of gas molecules. The increase in the thermal energy of the gas leads to the increasing rate of random collisions between the pendulum and the gas molecules, which is associated with an increase in noise. Similarly, an increase in an internal friction of a suspension fiber may only be due to an increase in the number of random atomic dislocations within the fiber, which corresponds to an increase in the uncontrolled random forces acting on the suspended test mass.

A more systematic exploration of this topic leads to the consideration of the fundamental *thermal noise*.

21.5   THERMAL NOISE

21.5.1   HISTORICAL OVERVIEW

The atoms and molecules which constitute any macroscopic object undergo thermal motion. In particular, thermal motion is present within any masses which would comprise the GW detector. Since the thermal motion is entropic by definition, it is inevitably a source of noise.

Such a noise is more fundamental than seismic noise. It produces a fundamental limit to any precision measurement in any domain. A closely related effect is the Brownian motion: the incessant chaotic motion of microscopic particles suspended in any liquid. Robert Brown first observed the effect in 1827, and the correct explanation that this motion is the result of impulses which suspended particles receive from the molecules of the surrounding liquid was given in the 1870s by Carbonelle and Delsaulx. Einstein's [26] and Smoluchowski's [27] theoretical investigations in 1905–1906 showed that the properties of Brownian motion are directly related to the viscosity, and paved the way for the fundamental *fluctuation-dissipation theorem* [28].

This limit was first recognized as far back as the early 1930s, when precise measurements of electric currents were being conducted.

21.5.2   THERMAL NOISE LIMIT

Let us, again, take a look at a simple harmonic oscillator, but from a different point of view. Any physical system in equilibrium has some finite temperature. What can we conclude about the motion of the SHO from the simple fact that its temperature is T?

According to the Equipartition Theorem, each degree of freedom of any system at temperature T possesses a random amount of energy equal, *on average $k_B T/2$*.

Assuming that the SHO is a mass-on-a-spring system, its mechanical energy is $m\omega_0^2/2$. Setting the average thermal energy equal to the mechanical energy leads to the conclusion that the oscillator must be randomly moving around the average root mean square displacement of magnitude

$$x_{rms} = \sqrt{\frac{k_B T}{m\omega_0^2}}. \tag{21.12}$$

This result shows that irrespective of the other properties of the system, it will have a mean square fluctuation, which can be reduced by either lowering the temperature, or by increasing the system's total mass (or both). The detector designers have pursued both of these pathways. The test masses of the current advanced interferometric detectors measure in tens of kg. The resonant mass detectors (which predate interferometric ones) reached masses of several metric tonnes. Most of the later stage resonant mass detectors were cryogenically cooled. Although the first detection of GWs was achieved by interferometers without cooling, most proposals for future detectors include cryogenic cooling.

21.5.3   THERMAL NOISE SPECTRUM

The thermally-induced root mean square displacement obtained above does not produce the noise spectrum. Moreover, it does not show any dependence on the material properties. This shortcoming is remedied by the fluctuation-dissipation theorem.

Let us again consider the simple harmonic oscillator, as we did in Eq. (21.2). This time, however, we include the dissipation term β:

$$m\frac{d^2x}{dt^2} + \beta\frac{dx}{dt} + k(x - A_{in}e^{i\omega t}) = 0. \tag{21.13}$$

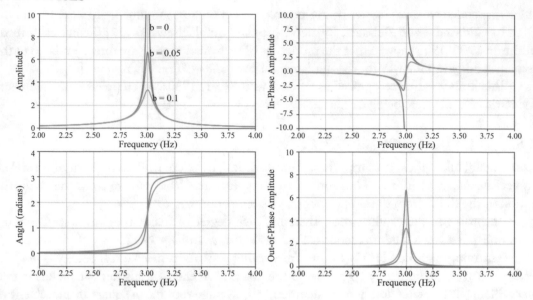

Figure 21.4: An example of the resonance curves of a lossy harmonic oscillator. The graphs on the left show the magnitude and phase, while the graphs on the right show the in-phase and out-of-phase amplitudes separately. Thermal noise is especially significant at frequencies where the out-of-phase amplitude (or, more precisely, in-phase velocity) is largest.

The solution of this equation is more complicated. An undamped system always oscillates either exactly in-phase or exactly anti-phase with the external disturbance, which follows from the fact that the transfer function in Eq. (21.4) is a real number. The transfer function is obtained for Eq. (21.13) by the same method as previously: assume that the induced motion has the same frequency as the external disturbance: $x(t) = A_{out}e^{i\omega t}$ and substitute. The result is a complex-valued transfer function

$$\frac{A_{out}}{A_{in}} = \frac{\omega_0^2}{\omega_0^2 - \omega^2 + i\beta\omega_0}. \tag{21.14}$$

The real and the imaginary parts of the transfer function can be plotted separately. When the frequency of the oscillator is close to the resonant frequency, the displacement described by the imaginary portion of the transfer function occurs out of phase with the driving force. This is the only motion present for the ideal oscillator without dissipation (at the exact resonance the amplitude of the motion of an ideal oscillator would increase without limits, which, of course, would mean the destruction of the device).

The presence of dissipation decreases the component of the motion in which the position of the mass is in-phase with the external disturbance. However, it creates the new component of motion, component, in which the position of the mass is out-of-phase with the disturbance.

Let us recall now, that in an oscillatory motion the *velocity* of the mass is out phase with the position. This means that if the *position* of the mass is out of phase with the force, then the *velocity* of the mass is *in phase* with the disturbance. The dissipation introduces the component of the motion in which the mass *moves* in phase with the applied force.

The fluctuation-dissipation theorem states that the effect of thermal motion is equivalent to a random force whose power spectrum is the same as the spectrum of this dissipative component of the motion.

To put this statement more precisely, we introduce the quantity called the *admittance* of an oscillator. This quantity is reminiscent of the transfer function, but instead of calculating the complex ratio of amplitudes of motion, it calculates the complex ratio of the velocity of an oscillator to the external driving force. If the system is driven with a force $F_{in} = F_0 e^{i\omega t}$, and, as the result, oscillates with velocity $v_{out} = v_0 e^{i(\omega t + \phi)}$, then the admittance is given by

$$Y(\omega) = \frac{v_{out}}{F_{in}} = \frac{v_0}{F_0} e^{i\phi}. \tag{21.15}$$

The fluctuation-dissipation theorem then states that the thermal noise is proportional to the real part of the admittance:

$$S_x(\omega) = \frac{4k_b T}{\omega} \operatorname{Re}(Y). \tag{21.16}$$

For the simplest case of SHO which we are discussing, the velocity is

$$v_{out} = \frac{dx}{dt} = \frac{d}{dt} A_{out} e^{i\omega t} = \frac{i\omega \omega^2 A_{in}}{\omega_0^2 - \omega^2 + i\beta\omega_0},$$

while the driving force is $F_{in} = k A_{in} e^{i\omega t}$. Dividing, we obtain

$$Y = \frac{i\omega}{m(\omega_0^2 - \omega^2 + i\beta\omega_0)}.$$

Calculating the real part, we, finally, can obtain the power spectrum of thermal noise of a damped SHO:

$$x^2(\omega) = \frac{4k_b T}{m\omega^2[(\omega_0^2 - \omega^2)^2 + \beta^2 \omega_0^2]}.$$

CHAPTER 22

Mechanical Detection

22.1 OVERVIEW

In the late 1950s through the early 1960s, J. Weber began research on detection of GWs by the means mechanical amplifiers, or mechanical GW antennas. Although his work resulted in controversy when he prematurely declared the detection, he was the founder of the field of experimental GW detection, which has continues lively research into the present. All attempts at the mechanical detection of GW have now been abandoned in favor of more sensitive interferometric detector, however a large number of the current techniques, both theoretical and experimental, were first discussed and tried using mechanical GW antennas. A brief overview of the ideas is in order.

As has been discussed previously, GR begins from the consideration of the effect of gravity upon free-falling TMs. The presence of gravity is impossible to recognize by the use of a single free-falling TM. A pair of TMs, on the other hand, undergoes tidal acceleration due to the tidal field. The idea of the mechanical resonators is to consider two TMs which are mechanically connected. The tidal fields then causes the "spaghettification" of the mechanical connection.

Originally, Weber proposed the use of a large piezoelectric crystal as the detector. The GW-induced stress within the entire crystal would then generate the electric potential difference between its ends "large enough to be observed with a low-noise radio receiver." However, the impossibility of obtaining a large block of piezoelectric material "as a single crystal" lead to the actual experimental work being done with a large block of metal. The stretching or compression of a mechanical object, such as a large metal bar, may be measured by a variety of methods. Weber's first practical scheme was to equip the bar with piezoelectric strain gauges, glued around the circumference.

The typical advanced bar detector was a 2–3 m long aluminum cylinders with the cross-sectional radius of 1 m, approaching 10^4 kg in mass: "Weber bars." The bars were suspended from a multi-stage vibration isolation stacks, designed to provide hundreds of dB of isolation at the bar's fundamental longitudinal resonant frequency near 1 kHz. The bars were located within a large cryostat, capable of reducing the bars' temperatures to sub-Kelvin range through the use of dilution refrigerator. Table 22.1 provides a list of the advanced bar detectors in the mid 1990s.

The central question is whether the stretching/compression of any mechanical connection produced by the GW may be amplified sufficiently to be measurable. In all experiments the mechanical connection was fashioned in such a way that the periodic stretching and compressions associated with the passing GW would set up resonant vibrations. These resonant vibrations,

Table 22.1: Advanced bar detectors in the mid-1990s

Beijing	Aluminum	Room temperature	Piezoelectric transducers
Guangzhou	Aluminum	Room temperature	Piezoelectric transducers
LSU, Louisiana	Aluminum	4 K	SC transducer with SQUID
MSU, Russia	Ultrahigh-O sapphire		Quantum nondemolition
Stanford	Aluminum	4 K	SC transducer with SQUID
Maryland	Aluminum	4 K and 300 K	SC transducer with SQUID
Rome	Aluminum	4 K	Electrostatic transducer
UWA, Australia	Niobium	4 K	Microwave with SQUID

rather than just the original stretching/compression, were amplified and then detected using piezoelectric sensors, accelerometers, or microphones.

22.2   TRANSDUCER DESIGN

Most research groups utilized *resonant transducers*, invented by H. J. Paik [34].

Let us treat the bar as a simple harmonic oscillator of spring constant K. The "mass at the end of the spring" for this oscillator is the mass of the bar itself, M. The frequency of the oscillator (in the actual experiments: the fundamental frequency of the bar) is given by the usual expression $\omega_0 = \sqrt{\frac{K}{M}}$. We need to detect the changes in lengths L of this oscillator.

A resonant transducer is another, smaller oscillator attached to the bar GW antenna. Its resonant frequency is chose to be the same as the frequency of the bar:

$$\sqrt{K/M} = \sqrt{k/m} \tag{22.1}$$

the *proof mass m* attached to the bar with a spring whose spring constant is k constitute the resonant transducer.

The motion of such a system of coupled oscillators has been discussed in the previous chapter. The resonant frequencies of such a system are give by Eq. (21.8). The mass and the spring constant of the transducer are much smaller than the the mass and the spring constant of the bar: $k << K$ and $m << M$. The resonant frequencies may be expanded in the terms of the small mass ratio $\alpha \equiv m/M$:

$$\omega_{21} \equiv \sqrt{k/M} = \alpha\omega_0, \tag{22.2}$$

so Eq. (21.8) reduces to

$$\omega_{0\pm}^2 = \omega_0^2 \left(1 \pm \frac{\sqrt{\alpha}}{2}\right). \tag{22.3}$$

The superposition of these two nearly identical frequencies produces *beats* between the modes.

The ratio of the amplitudes of motion of the two masses m and M then can be obtained as

$$\frac{x}{X} = \mp \frac{2}{\sqrt{\alpha}}. \tag{22.4}$$

The oscillation of the proof mass is thus a factor $2/\sqrt{\alpha}$ larger than the motion of the bar.

22.3   THERMAL NOISE REQUIREMENTS

Consider, first, the size of random variations in the length of this oscillator due to the thermal motion of its constituent atoms. Using Eq. (21.12), we obtain for the rms motion of its end

$$\frac{\Delta x_{thermal}}{L} \propto h_{thermal} \propto \frac{1}{L}\sqrt{\frac{k_B T}{M\omega_0^2}}. \tag{22.5}$$

Plugging in the numbers quoted above we find $h_{thermal} \approx 4 \times 10^{-18}$ at the lowest achievable temperatures (10 mK). This thermal RMS strain motion increases with the square root of the absolute temperature.

Fortunately Weber bars have much better sensitivity than the above calculation suggests. The bars are designed with high mechanical quality factor Q. The quality factor of an oscillatory system describes the degree of damping:

$$Q \equiv \frac{\omega_0}{2\beta}, \tag{22.6}$$

where β is the damping parameter. The values of Q vary greatly in ordinary physical situations. The quality factor for a loudspeaker, for instance, may very from less than ten to a hundred. Tuning forks may have Qs of 10^4. Typically, higher values of the quality factor are observed in non-mechanical oscillatory objects. Electromagnetic resonant cavities may have $Q \approx 10^5$, and the Q associated with an atomic resonance may be on the order 10^7.

The material for a Weber bar is chosen so as to guarantee very high mechanical quality factors at the working conditions. As an example the mechanical quality factor for AURIGA bar is about 4×10^6 at 100 mK.

The state of vibration of an object with a high Q changes very slowly in response to a stationary noise force. Under the influence of thermal excitations, the time it takes for the oscillator to "forget" its previous amplitude or phase, the *decoherence time*, is on the order of $\tau_d \propto Q/f_0$. The magnitude of the thermal noise reaches the above estimates (above 10^{18} for sub-kelvin temperatures) only at the timescales long compared to τ_d. If the measurement time τ_m is short compared to τ_d the effect of the thermal noise is reduced by a factor $\sqrt{\tau_m/\tau_d}$. In other words, the long "ringing" of the bar would allow the detector to "average out" the thermal noise.

A figure of merit in pulse detection is the *pulse detection noise temperature T_p*, defined as the amount of energy that must be deposited by a pulse to achieve signal-to-noise ratio $SNR = 1$.

For the bar limited only by thermal noise it is

$$T_p = \frac{T}{Q}\tau_m\omega_0. \tag{22.7}$$

22.3.1 SENSOR NOISE

Thus to overcome thermal noise we need long measurement times. Unfortunately, longer measurement times decrease the performance due to *sensor noise*: the noise from the transducer and amplifier themselves. A large fraction of effort in the era of resonant bar GW detectors went into improving these noises.

CHAPTER 23

Interferometry

Optical interferometry has served as the most precise method of measuring small variations in distance since at least 19th century. It has played a major role in bringing down the aether theory and paving the road for special relativity. Over a hundred years later it provided the means for the first direct observation of GWs. Several proposed methods of GW detection rely on some modification of interferometric methods. Before discussing those methods, let us review the general principle.

An interferometer measures the variation in optical path length by comparing the phases at different locations of the same EM wave train. Let us consider a perfectly coherent plane EM wave. Such a wave is described by a perfect sinusoid, with electric given by

$$\mathbf{E}_0 = \mathbf{E}_{\max} \cos(\mathbf{k} \cdot \mathbf{r} - \omega t + \phi_0), \tag{23.1}$$

where $\mathbf{E}_{\max}$ is a vector of constant magnitude and direction. When this EM wave is sent into an optical detector, the detector registers the *irradiance* of the wave—the average energy per unit area per unit time. More commonly called *beam intensity*, it is proportional to the square of the amplitude of the electric field, $I \equiv \frac{c\epsilon_0}{2} E_{\max}^2$. To perform an interferometric measurement, a "copy" of this wave is obtained by some means, and is made incident on the same detector. If the "copy" arrives to the detector with exactly the same phase as the "original," the magnitudes of the electric fields of the two waves add up to produce twice the electric field. Since the intensity is given by the square of the electric field amplitude, this results in *four* times the original intensity.

Exactly the same result is obtained if the EM "copy" wave is shifted with respect to the original one by one full wavelength. However, if the "copy" is shifted by one-*half* the wavelength then the two electric fields perfectly cancel each other, and the resulting intensity is zero.

If the detector registers the intensity between zero and the maximal, we can conclude that the "copy" is shifted by less than one-half the wavelength, and calculate this shift.

The best developed laser for GW research is the Neodium-YAG laser which produces near infrared light at 1064 nm. Traditional interferometry with this laser is, thus, capable of measuring the distance variations below 512 nm.

This value—half-a-micrometer—is at once too large and too small. It is too large because the distance variations produced by GW on Earth are many orders of magnitude below this value. So one of the questions that have to be answered is: how much smaller than half-a-wavelength of light can interferometry measure? In other words, what is the *sensitivity* of the interferometer?

On the other hand, this value—half-a-micrometer—is too small for our ability to build a macroscopic detector with this precision. If interferometry requires that the "copy" of the wavetrain is shifted by no more than half a micrometer, then the interferometer itself would have to be no larger than, perhaps, a few millimeters in size (otherwise it would not be possible to recombine the waves with sufficient precision). So the second question that has to be answered is: by how many additional wavelengths of light can the "copy" be separated so that variations in distance are still measurable.

The answers to these questions determine some of the noises inherent in the process of using light to measure distance. The *shot noise* is the fundamental noise that sets the lowest limit on our ability to measure ever smaller distances. The *radiation pressure noise* is the fundamental noise that (predictably) limits the amplitudes of the electric fields in the EM wave which we can use for the measurements. Finally, the *laser phase noise* determines how far the two "copies" of a laser light wave can be shifted with respect to one another. The various methods of evading phase noise drive the designs of different detectors.

23.1   SHOT NOISE

If two copies of the same EM wave are in phase when they are incident on an optical detector, their electric field magnitudes add up to produce four times the intensity of the single original EM wave—the maximum brightness that can be obtained from the two copies. If these two copies are, instead, shifted by half-phase, the detected power will be zero. When the two copies of the same EM wave are shifted by some different fraction of the wavelength of light, the resulting intensity at the detector is somewhere between the maximal and zero. How sensitive the interferometer is to the changes in displacement then comes down to the question of what smallest gradations of "brightness" are possible.

The ultimate limit derives from the quantum nature of light. The light energy is absorbed in light quanta: photons, each of which carries $h\nu = hc/\lambda$ units of energy, less than 10^{-18} J for visible light. An extremely naive expectation would be that these energy quanta may set the lowest limit of sensitivity. One could imagine a 1-W laser producing a uniform steady stream of 10^{18} photons per second. Perhaps we can, in principle, detect the relative differences in brightness corresponding to a loss or a gain of one photon out of 10^{18} per $\Delta t = 1$ second?

There are several reasons why this picture is too optimistic. The central reason is that the photons do not arrive in a uniform steady stream. The quantity

$$\Phi = \frac{P}{h\nu} = \frac{\lambda P}{hc} \tag{23.2}$$

does describe the *mean photon flux*, and, *on average*, in time Δt the number of photon arriving at the photodetector is given by $N_{ave} = \Phi\delta t$. However, the *actual* number of photons arriving at the detector varies randomly, with some *probability distribution* describing how many photons *might* arrive at the detector within any given time Δt. If a GW causes a variation in brightness which corresponds to a single photon, that variation will be "lost" in larger brightness variations caused by the randomness of photon arrival.

The *standard deviation* σ of a probability distribution is used to describe the magnitude of the average deviations of the actual from the mean. To conclude that a certain variation in the amount of power arriving at the photodetector is due to anything other than the randomness inherent in photon arrival times, that variation has to be a certain number of standard deviations above the variation due to the photon counting statistics.

These natural variations in brightness should be interpreted as a specific form of noise, the *photon shot noise*. To quantify this noise, we calculate the distance by which two sinusoid wavetrains would have to be separated to produce the change in brightness equivalent to one standard deviation of shot noise.

The magnitude of the standard deviation depends on the specific probability distribution associated with photon statistics. That probability distribution depends on the nature of the light source. For a coherent monochromatic plane wave it is the *Poisson probability distribution*. For Poisson distribution, one standard deviation is equal to the square root of the mean:

$$\sigma_{Poisson} = \sqrt{N_{ave}},\tag{23.3}$$

therefore the standard deviation of the number of photons arriving at the photodetector in time $\Delta t = 1$ s is

$$\sigma_{\text{photons in }\Delta t} = \sqrt{\Phi \Delta t} = \sqrt{\frac{\lambda P \delta t}{hc}}.\tag{23.4}$$

This means that the "extremely naive" sensitivity of 1 photon out of 10^{18} gets reduced to only $\sqrt{10^{18}}$ photons out of 10^{18}. A perfect interferometer built with ultra-stable 1-W laser and a 100% efficient photodetector is then capable of measuring distance variations of 10^{-9}th of half-wavelength, or about one-half of a femtometer ($1\text{fm} \equiv 10^{-15}$ m). This is impressive, especially considering that the diameter of an atomic *nucleus* is on the order of a femtometer, however it is not sufficiently small for the detection of a GW.

23.2   RADIATION PRESSURE NOISE

If the "extremely naive" picture of an ordered string of photons producing a EM wave of constant power were correct, the sensitivity of the interferometer could be increased by increasing the power. So, if 10^{18} photons pass through the 1-W beam in a second, then 10^{19} photons pass through a 10-W. We could get an increase in sensitivity proportional to the increase in power.

At first glance, the more realistic picture of random arrival of photons still appears to provide more sensitivity with increasing power, though not quite as much. Instead of sensitivity of $1/P_{ave}$, we have to satisfy ourselves with the sensitivity of $\sqrt{P_{ave}}/P_{ave} = 1/\sqrt{P_{ave}}$. There is an increase in sensitivity proportional to the square root of the increase in power!

Unfortunately, the randomness of photon arrival creates another fundamental type of noise, called *radiation pressure noise*, and this type of noise increases with power. This time the culprit is not the light arriving at the detector, but the light incident on any kind of mirror, beamsplitter, or on any other hardware that is used to alter the path of the EM wave.

EM waves exert *radiation pressure* on the surfaces upon which they are incident. The magnitude of the pressure is equal to the energy density of the wave, which can be expressed as

$$\text{Pressure} = \frac{P}{cA}, \tag{23.5}$$

where A is the surface area over which the EM wave is incident. Since the power P effectively fluctuates with the random arrival rate of the photons, the pressure fluctuates as well. This inevitably produces random variations in the separation of the two copies of the EM wave trains used to measure the distance.

To evaluate the effects of the fluctuating radiation pressure, suppose that the optical element responsible for guiding the beam has mass m. The fluctuating force exerted on this mass is $F_{rad} = \text{Pressure} \times A = P/c$. he average force is, presumably, somehow compensated the mounting of the optical element. The force fluctuation σ_F, however, produces an "unexpected" position change in the amount

$$\sigma_x = \frac{1}{2} \frac{\sigma_F}{m} (\Delta t)^2 \tag{23.6}$$

over the small amount of time Δt.

The magnitude of force fluctuations follow directly from the fluctuations in the number of photons Eq. (23.3), so:

$$\sigma_F = \sqrt{\frac{P}{c^2 h\nu}} \tag{23.7}$$

and

$$\sigma_x = \frac{\Delta t}{m} \sqrt{\frac{P}{c^2 h\nu}}. \tag{23.8}$$

23.3   PHASE NOISE

The brightness at the output of a plane-wave interferometer varies between the maximum and the minimum as the two copies of the same EM wavetrain are shifted by less than $\lambda/2$. What if the two wavetrains are shifted by a larger distance?

For a pair of perfect plane sine waves, the answer is simple: the variations in the intensity of the sum of the two waves will oscillate. The intensity will keep reaching the maximum whenever the two copies are separated by an integer number of wavelengths $n\lambda$. It will fall down to zero whenever the two copies are separated by $(n + 1/2)\lambda$. This means, that it should not be very difficult to construct an interferometer that measures small distance variations: the two copies of the sine wave may be separated by an arbitrary distance.

However no EM source is capable of producing perfect sine wave. Apart from the quantum mechanics related randomness of photon arrival, there are also inevitable deviations from

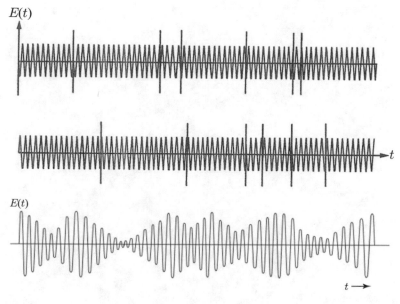

Figure 23.1: A crude representation of a quasimonochromatic wavetrain. The top two lines attempt to show that the phase of the light emitted at one moment differs from the phase of the light emitted at a later moment. The bottom line includes also the effects of random amplitude fluctuations. Both effects are grossly exaggerated.

the perfection of mathematical sine wave. In particular, any EM wave source possesses an attribute called *coherence length*, which describes "the extend in space over which the wave is nicely sinusoidal so that its phase can be predicted reliably" [29] (Fig. 23.1).

This leads to the *phase noise*. The methods of dealing with this noise lead to the design of various types of detectors discussed in several of the upcoming chapters.

CHAPTER 24

Terrestrial Interferometric Detection

24.1 MICHELSON INTERFEROMETERS

The direct observation of GWs has resulted from the use of terrestrial Michelson laser interferometers with Fabri–Perot cavities in the arms.

Let us first consider a simple Michelson interferometer whose components are floating freely in space, at rest with each other, and far removed from any gravitating bodies (Fig. 24.1). It consists of a light source L, a beam splitter BS (that is, a partially reflecting mirror), and a pair of mirrors MX and MY some distance from the the BS. The BS divides the beam from the source into two, one (e.g., transmitted) portion traveling along the X-arm, the other along the Y-arm.

In the X-arm the light, transmitted by the BS, is then reflected by MX back toward the BS. Part of this returning light is now reflected by the BS toward an optical detector. (The other part of the returning light is transmitted again by the BS. This doubly transmitted light then travels back toward the source. If this light were not captured and re-routed, it would produce undesired loss of power and/or interfere with the workings of the source itself.)

The Y-arm receives the light reflected by the BS. The MY mirror reflects the light back toward the BS, which transmits part of this returning light toward the optical detector. (The other part of the returning light is reflected again by the BS toward the the source, and has to be re-routed.)

When the waves from the X- and from the Y-arm are recombined after the BS (both at the optical detector, and at the source), the interference is to be expected. The relative optical lengths of the two arms may be manipulated to produce either destructive, or constructive interference at the detector. More generally, by an appropriate manipulation of the relative optical lengths of the two arms, any desired phase shift between the light in the two arms may be obtained.

The optical length may be manipulated either by altering the index of refraction in one of the arms, or, simply, by altering relative physical lengths of the arms. When the light source is not monochromatic, a compensation plate is necessary to overcome such practical issues as light dispersion. The light sources of interferometric GW detectors are highly stabilized lasers, however most GW detectors do use some version of phase compensation.

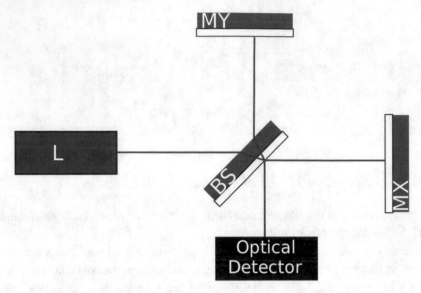

Figure 24.1: A basic Michelson interferometer.

24.2   A GRAVITATIONAL WAVE AND A MICHELSON INTERFEROMETER

Consider now the effect of incoming GWs. A plane GW, propagating perpendicularly to the plane of the interferometer, with "+" polarization aligned with the two arms, alternately accelerates the end mirror in one arm toward the beam splitter, while simultaneously accelerating the end mirror in the other arm away from the beam splitter. This leads to a change in the relative arm lengths. The light entering the interferometer experiences a phase shift, which is measured at the detector.

From the early times of experimental GW science an objection to this method of GW measurement has often been voiced. This objection is sometimes termed "The rubber ruler objection," and goes as follows: "The light waves cannot be used to measure the change in the distances between the mirrors, since the GW stretches and contracts the light waves themselves." More precisely: if the GW affects the wavelengths of the light in the arms of the interferometer in the same proportion as it affects the lengths of the arms themselves, then the relative phase of the two light waves is not affected. Therefore, we should not expect to observe a change in the interference pattern.

There are several equivalent explanations of why such an objection is not valid. The simplest one is that the light which is used to "measure" the changed distance between the mirrors is not the same light that was present within the interferometer before the passing of the GW. The terrestrial interferometers are attuned to the GW frequencies of around 100 Hz, so it takes on the order of $1/(100 \text{ Hz}) = 10^{-2}$ s for a single GW wavelength to pass. Compare this to the amount of time the light spends in the interferometer. The arms lengths of largest present day interferometers (LIGO) equal 4 km, so the single round-trip time from the input to the detector is less than 3×10^{-6} s. In actuality, the interferometers are constructed so that the light spends about 1000 times as long inside the interferometer. Still, this means that the "new" light enters the interferometer while the length of the arms are affected by the GW.

In other words, the GW detector interferometer does not use the instantaneous separations between the wave crests of the light as the "marks on the ruler." Instead, it effectively measures the time that it takes for a given wave crest to travel the length of an arm. The presence of the GW alters the spacetime curvature,

$$(1 + h_{11}(\omega_{GW} t - \mathbf{k} \cdot \mathbf{x})) dx^2 - c^2 dt^2 = 0,$$

so the distance between the spatial coordinates of two free-falling points changes from dx to $\sqrt{1 + h_{11}} dx$.

More generally, if the direction of the GW propagation make angle θ with the direction of the laser light beam, then the arrival time of a given wave crest will be altered by

$$\frac{dt_f}{dt} = 1 + \frac{1}{2}(1 + \cos\theta)\{h_+[t + (1 - \cos\theta)L] - h_+(t)\}. \tag{24.1}$$

24.3   HISTORY AND DESIGN OF ACTUAL INTERFEROMETRIC GW DETECTORS

The first detection of a GW on 14 September 2015 [30] has earned a Nobel prize to three scientists, members of LIGO-Virgo collaboration. One half of the prize was awarded to Rainer Weiss, who was the "mastermind" behind the entire LIGO enterprise. Theoretical physicist Kip Thorne, the leading figure in the modern understanding of gravitational physics, and of GW generation in particular, and B. Barish, the scientific leader who brought the project to completion shared the other half of the prize.

The first discussion of the possibility of using interferometry for GW research appeared in a 1962 paper by Soviet physicists M. Gertsenshtein and V. Pustovoit. Although that work was correct in many of its assumptions and conclusions, it was not followed up by any experimental research. The true birth of the interferometric GW detection occurred independently in the USA in the early 1970s. In 1971, Moss, Miller, and R. Forward published a paper describing the development of "long wideband" GW detector that started at Hughes Research Lab in 1966. The authors credited the original idea to P. Chapman of NASA, and to the discussion with R. Weiss. This first interferometer consisted of flat mirrors rigidly attached to an optical table, with one of the mirrors mounted on a piezoelectric element. The last paragraph of the paper listed proposals for building a suspended interferometer with heavy mirrors, placed in a vacuum system and extended in size to several kilometers.

In 1972 R. Weiss produced a research progress report titled "Electromagnetically Coupled Broadband Gravitational Antenna." This report contains the first comprehensive design of an interferometric GW detector. It contains a lucid discussion of most relevant requirements, and discusses most of the noise sources that the present day GW researchers encounter every day.

After thirty years of development, the first kilometer-scale interferometers started operation in the early 2000s. At the time of the first detections there were four such systems: two identical **Laser Interferometer Gravitational-Wave Observatory** (LIGO) interferometers located in the opposite corners of the US (Hanford WA, and Livingston LA), a single Virgo interferometer located in Cascina, Italy, and GEO600 located near Hannover, Germany. Virgo interferometer was not operational during the first detections. GEO600 is used for active research into advanced methods of GW detection, but is not expected to be sensitive to most GWs. The first detections were done by LIGO.

All present day interferometers are Michelson interferometers whose arms are *Fabry-Perot cavities*. Having the cavities in the arms enables the light to "stay" inside the interferometer hundreds times longer than would be possible in a single-pass Michelson. It also multiplies hundredfold the available power, greatly reducing the shot noise.

The light source for the interferometers is a 35 Watt ultra-stable laser with operating wavelength 1064 nm (additional amplifiers can extend the power of the laser up to 200 W). In addition to the main interferometer the system contains a number additional suspended optical systems: input and output *mode cleaners*, which perfect the shape of the beams before and after the main interferometer and stabilize the frequency of light, and a power recycling cavity, which return a portion of the light reflected by the main interferometer.

The test masses of the GW detectors are the mirrors of the interferometers. Each test mass is a 40-kg fused silica glass substrate. A LIGO test mass is the last stage of a four-stage pendulum. These pendulums themselves are installed on additional seismic-isolating stages in ultrahigh vacuum environment. Each arm of the LIGO interferometers is four kilometer long.

The Virgo interferometers have three kilometer arms. The test masses are suspended from 7.3-m long structures called *superattenuators*. A superattenuator is a chain of pendula, supported by three inverted pendula.

Another interferometric GW detector under development is **Kamioka Gravitational Wave Detector** (KAGRA) operated by the University of Tokyo. It is the first underground GW detector, and the first detector with cryogenic mirrors.

CHAPTER 25

Space-Based Detection

Terrestrial detectors are capable of detecting the GW having the frequencies in the range of 100–1000 Hz. The GW in this range of frequencies are produced by objects with masses a few 10's that of the Sun. Sources with much larger masses, such as the mergers of super-massive black holes produce signals at much lower frequencies. The detection of low-frequency signals requires much larger detectors.

The ideal location for very large detectors is space. *Light Interferometry Space Antenna* (LISA) is the proposed network of satellites exchanging laser light (Fig. 25.1). The laser light sources and mirrors which are installed on the satellites, compose a Michelson interferometer in space with an arm length of 2.5 million km. The arm length allows observations of most of the GWs in the range of frequencies from 0.1 mHz to 100 mHz. Such GWs should originate from many super massive BH sources.

25.1 SPACECRAFT POSITIONING

The LISA mission consists of three spacecraft, each orbiting the Sun in its own independent elliptical orbit. It is not possible to have them remain at exactly the same initially chosen distances from each other. However, the three independent orbits are so chosen that the variations in separation are minimized. With this careful choice of orbits, the spacecraft always remain at the vertices of a near-equilateral triangle without the need for orbital corrections. More precisely, the *guiding center* of the formation (geometrically, the guiding center is the circumcenter of the triangle formed by the craft, or the point equidistant from all three) is in the ecliptic plane 1 Astronomical Unit (AU) from the Sun and 20° behind the Earth (about 50 million km).

The plane of the triangle is inclined by 60° with respect to the ecliptic (this is the angle which minimizes the variations in separations [31]). The configuration remains stable throughout the year, with the triangle appearing to rotate about the center of the formation (Fig. 25.2).

As with terrestrial interferometric observatories, a interferometric space antenna aims to detect the passage of a GW by measuring the time-varying changes of optical path length between free-falling test masses. The arms of the interferometer are defined by these TMs, and the spacecraft are, essentially, built to "float" around TMs. Each spacecraft contain lasers, the interferometry equipment, and capacitive sensors for measuring its position with respect to the TM inside of itself. A control system ("Drag Free Attitude Control System") commands the micro-Newton thrusters of the spacecraft to adjusts itself to the TMs if needed. This produces

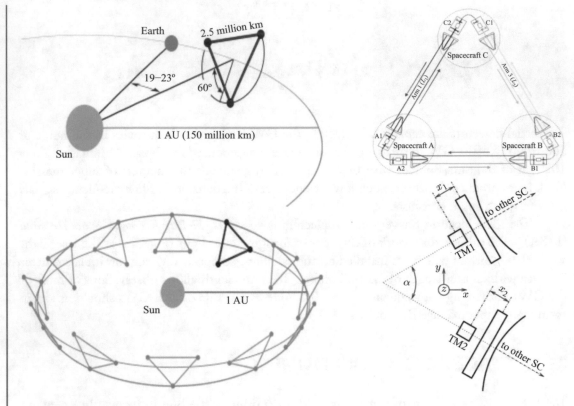

Figure 25.1: LISA.

drag-free operation. Drag-free operation reduces the disturbances to the test masses caused by non-gravitational forces affecting the spacecraft.

The mean separation between the TMs is 2.5 million km, or 8.4 light seconds. Because the TMs are in the independent orbits, they move with respect to each other with velocities of about 5 m/s. The distances between them (the lengths of the "sides" of the almost equilaterla triangle) change by as much as 50 thousand km (0.16 light seconds) over the course of the year.

25.2   INTERFEROMETRY EQUIPMENT

Each craft contains interferometry equipment for measuring its position with respect to the other craft (Fig. 25.3). Thus the measurement of the changes in the arm length consists of three distinct parts: one is the measurement of the distance between the spacecraft, two other are the measurements between each of the TMs and its respective optical bench.

A laser beam is sent out from each spacecraft to each one of the other two. The beam diverges over several million kilometers, so the amount of light power received from the other

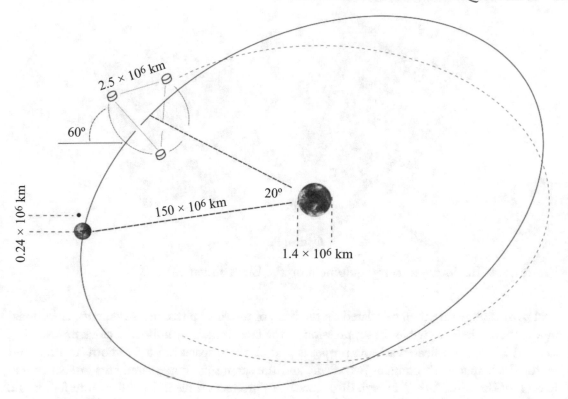

Figure 25.2: LISA constellation. The distances are not shown to scale. The mean Earth-Moon distance of 0.3844 million km. The orbit of each TM is independent of the orbits of the other TM. One such orbit is shown.

crafts is limited to $100pW$. It is therefore not possible to use a normal Michelson interferometer for the measurement of the spacecraft-to-spacecraft distances, or even to rely on passive reflection. Instead of the mirrors, the receiving spacecraft are equipped with their own lasers. The lasers in the receiving spacecraft are phase-locked to the incoming beams, and send fresh high-power beams back.

In the simplest scheme, one of the three spacecraft is "central," and the other two are "end" spacecraft. The central craft is built around two TMs, each of which is dedicated to a separate arm of the interferometer. A laser beam is sent from the "center" to each of the two "ends." The lasers at the two "ends" are phase-locked to the incoming signal. These end-lasers send fresh high-power beams back to the central craft. The central spacecraft receives returned signals, and compared their phases with the phases of its own local lasers. The phases of the two lasers on the central spacecraft are also compared within the central spacecraft. The relative op-

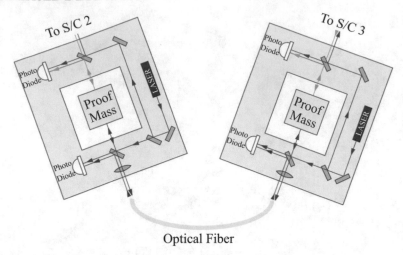

Figure 25.3: The interferometry equipment on the LISA aircraft.

tical path changes are then calculated on the basis of the three phase measurements. Additional modulation schemes enable the suppression of the laser frequency noise and clock noises.

A more complicated scheme proposes that all three spacecraft act as both "central" and "end." Each spacecraft contains two TMs, and the same sets of measurements are performed in each of the three "arms" in both directions. This produces three independent interferometric combinations: two Michelson interferometers, and one Sagnac interferometer. The Sagnac interferometer is not sensitive to tidal waves, and can be used to subtract unwanted background. Two independent Michelson interferometers can be used to measure two possible GW polarizations simultaneously.

The system is designed to allow the detection of strain below 4×10^{-24} in a 2-year measurement.

25.3   TIME DELAY INTERFEROMETRY

The separation between the pairs of craft cannot not be nearly identical most of the time. Therefore, unlike in the case of terrestrial GW detectors, the lengths of the "arms" of spatial interferometer cannot always be nearly equal. Therefore the exact cancellation of the laser phase noise does not happen. Let the laser frequency fluctuation be denoted as $C(t)$. If L_1 and L_2 are the lengths of the two arms, the relative frequency fluctuations after the beams are combined is

$$\Delta C(t) = C(t - 2L_1/c) - C(t - 2L_2/c). \tag{25.1}$$

The basic scheme for resolving this problem can be understood as follows. The beams are not made to interfere at a common photodetector directly. Instead, two separate photodetectors are used to record the interference between the light arriving from each arm, and the "new" light from the laser that is just being injected into the arms. The records of these data $y_1(t)$ and $y_2(t)$ have contributions:

- laser noise on the returning light $C(t - 2L_i/c)$,

- laser noise of the "new" light $C(t)$,

- possible GW signal $h_i(t)$,

- possible additional unknown noise $n_i(t)$.

The data streams obtained on the two detectors are

$$y_1(t) = C(t - 2L_1/c) - C(t) + h_1(t) + n_1(t)$$
$$y_2(t) = C(t - 2L_2/c) - C(t) + h_2(t) + n_2(t). \tag{25.2}$$

The two data streams are recorded, and can now be manipulated (added, subtracted, time-shifted) at will. The following manipulations then succeed at eliminating laser frequency noise:

(1) subtract the data streams:

$$y_1(t) - y_2(t) = C(t - 2L_1/c - C(t - 2L_2/c) + h_1(t) - h_2(t) + n_1(t) - n_2(t); \tag{25.3}$$

(2) time-shift the data stream $y_1(t)$ by the round-trip time in arm 2, do the opposite with $y_2(t)$:

$$
\begin{aligned}
y_1(t - 2L_2/c) &= C(t - 2L_1/c - 2L_2/c) - C(t - 2L_2/c) \\
&\quad + h_1(t - 2L_2/c) + n_1(t - 2L_2/c) \\
y_2(t - 2L_1/c) &= C(t - 2L_1/c - 2L_2/c) - C(t - 2L_1/c) \\
&\quad + h_1(t - 2L_1/c) + n_2(t - 2L_1/c);
\end{aligned} \tag{25.4}
$$

(3) subtract the time-shifted data streams:

$$
\begin{aligned}
y_1(t - 2L_2/c) - y_2(t - 2L_1/c) &= C(t - 2L_1/c) - C(t - 2L_2/c) \\
&\quad + h_1(t - 2L_2/c) - h_2(t - 2L_1/c) \\
&\quad + n_1(t - 2L_2/c) - n_2(t - 2L_1/c).
\end{aligned} \tag{25.5}
$$

The important point is that the way the laser phase noise enters in this equation, $C(t - 2L_1/c) - C(t - 2L_2/c)$, is identical with the way it enters Eq. (25.3) in Step (1) above. Subtracting the result of Step (3) from the result of Step (1), we produce the final data stream X:

(4)

$$X \equiv [y_1(t) - y_2(t)] - [y_1(t - 2L_2/c) - y_2(t - 2L_1/c)] = [h_1(t) - h_1(t - 2L_2/c)]$$
$$- [h_2(t) - h_2(t - 2L_1/c)] + [n_1(t) - n_1(t - 2L_2/c)] - [n_2(t) - n_2(t - 2L_1/c)].$$
$$(25.6)$$

Data stream X still contains the signatures of the GW strain, but laser phase noise has canceled exactly. This is the basis of the *time delay interferometry*, which, in more sophisticated form, will be utilized in space-based GW detectors.

CHAPTER 26

Atomic Gravitational Wave Detectors

26.1 ATOMIC, MOLECULAR, AND OPTICAL PHYSICS

Atomic, Molecular, and Optical (AMO) physics is a branch of physics with a venerable history. At the end of the 19th century it provided experimental knowledge that would eventually lead to the development of quantum mechanics. It entered a dormant period in the 1930s as nuclear physics grew in importance, but the invention of lasers in 1960 gave it a new impetus. The incremental increase in precision and stability of lasers culminated in the development of a host of new techniques in the 1990s. These again brought AMO to the forefront of physics research, in both the academic and industrial settings, capable of producing answers in the most fundamental branches of physics, as well as in very practical fields related to biological and even geological research.

The expansion of AMO physics during the 1990s was due to the development of various methods of trapping small assemblies of individual quantum particle—from several billion atoms down to single electrons. Prior to that time quantum mechanics could be thought to be mostly statistical in nature. In fact, the sentiment that it is not possible to perform real-world experiments with individual isolated quantum systems was expressed by several of the "fathers" of quantum mechanics. The advent of the quantum particle trapping techniques proved that the rules of quantum mechanics are probabilistic, but *not* merely statistical in nature. More importantly for the GW research, it provided the capabilities for *quantum state engineering*, whereby individual quantum systems may be placed into well-defined quantum states, or even into coherent superpositions of several well-defined quantum states, may be left in those well-defined quantum states for extended time durations, and may be transitioned between well-defined quantum states in a precisely controlled manner.

Quantum state engineering offers a new methods of GW detection. The central principle remains the same as in many other detectors: use the constancy of the speed of light to measure the distance between free-falling test masses. The difference is that the free-falling masses are now chosen to be microscopic objects, whose quantum states can be affected by, or even *entangled* with the incident light. The quantum state of the TM may be affected by a properly chosen *pulse* of light, and the same pulse of light can then travel to the next TM to modify its quantum state in turn (Fig. 26.1).

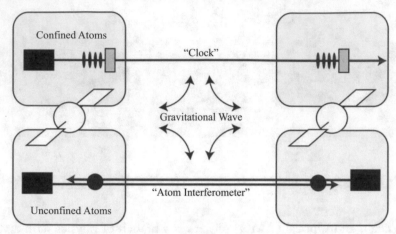

Figure 26.1: Atomic GW detectors. Adopted from [32].

The key feature of the interaction light-atom interaction is the ability to "write" the phase of the light onto the atom. Without going into the explanation of how, suppose that it can be done. Let us understand how this can allow for the novel type of measurement of the change in distance.

26.2   THE LAYOUT OF THE ATOMIC GW DETECTOR

As described in the Chapter 23, an optical interferometer uses the phase shifts of an EM wave to detect small changes in the distance between two points. In principle, the distance between the two points could be arbitrarily large, if only the EM wave were perfectly sinusoid. Unfortunately, the EM wave is never perfectly sinusoid, but rather is subject to phase noise. Phase noise limits the range of distances whose variations can be measured. To overcome this limitation one needs to devise a scheme (e.g., Michelson interferometer) to compare copies of the same wave which are not separated from one another by too many wavelengths.

Now, consider two atoms, a and b, separated by distance L. The GW modifies distance L as discussed in the previous chapters. The goal of the detector is to measure the change in this distance.

Suppose the laser source produces a pulse which "imprints" its local phase on atom a. This same laser pulse then propagates along long distances to atom b, and "imprints" its local phase on that atom. Laser phase noise does not contribute to the local phase variations of the pulse which has already been emitted. Both atoms received the signal from the same laser pulse, emitted by the laser earlier. Thus the laser phase noise does not create any phase difference between the phases imprinted on the atoms. The phase difference between the two recorded phases can result *only* from what happened to the laser pulse en route from atom a to atom b, e.g., from the variations in distance L.

In other words, the atoms act as a coherent phase memory that keeps track of the phase imprints of a laser pulse [32].

CHAPTER 27

Pulsar Timing Array

As was described in 15, neutron star pulsars produce extremely regular periodic flashes of observable radiation. More precisely, the periods very slowly increase as the stars lose their rotational energy due to a variety of processes. However, the rate of the loss is also extremely stable, and the period of pulsar flashes is extremely predictable.

Small changes in the regularity of these flashes indicate astronomical disturbance somewhere in the interstellar medium along the line of sight between the neutron star and the detector. A number of such disturbances are well understood and accounted for. These effects include the dispersion of the pulses as they encounter charged particles, the variations in the observed intensity known as scintillation, and the scattering of pulses by the irregularities in the interstellar medium.

A GW crossing the pulsar-detector line-of-sight is another possible disturbance, and should affect the time of arrival of the pulses. A pulsar timing experiment measures the *timing residual*: the difference between the expected arrival of the next pulse, and the actual time of arrival.

The effect of the passing GW on the travel time of light is described, as before, by Eq. (24.1). Since GWs are not the only phenomena affecting the arrival time of the pulsar radiation to the detector, it is insufficient to study the timing residuals of the radiation from a single pulsar.

A *pulsar timing array* is a set of several pulsars whose timing residuals are simultaneously measured (Fig. 27.1). If a single GW passes the line of sight from all pulsars in the array, a correlation in the timing residuals will be observed. This is not true of other phenomena affecting the timing residuals.

Obviously the implication here is that the GW is correlated over astronomical distances. The locations where the same GW crosses the lines of sight from two different pulsars are likely to be separated by distances of light years. The technique is therefore most applicable to detecting the GWs whose periods are measured in years, i.e., have frequencies in the several nHz range. The events that are capable of producing GW of such frequency are the interactions between slowly moving supermassive black holes at the centers of galaxies. These are expected to form binaries as a result of galaxy mergers.

The fact that the periods of these GWs are measured in years also implies that this method of detection requires years of observation. The changes in timing residuals on the order of a few nanoseconds have to be detected.

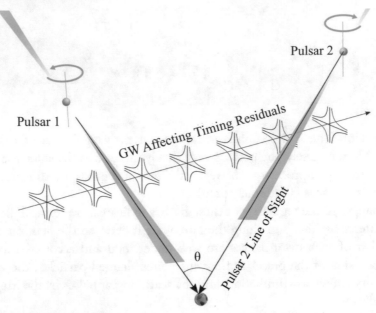

Figure 27.1: A Pulsar Array. In reality, there is not a single GW, but a stochastic field of many GWs propagating in different directions with wavelengths and polarizations.

To add a complication, there are always many different GWs within the Earth-pulsar line of sight. These GWs are produced by different sources, and they have different wavelengths, amplitudes, and polarizations. Each timing residual is affected by all of those passing GWs, so the measured signal arises from an unknown multitude of GWs. In other words, the pulsar timing arrays are involved in the search of stochastic GWs, which arise from a superposition of incoherent sources. There is no specific signal signature that may be searched for. The situation has been described as "trying to detect noise in noise." Fortunately, even such a random field of GWs is detectable by the use of *correlations* in the response of a system of detectors.

Although GWs are not the only cause of possible correlations, the correlations produced by GWs have a specific functional dependence which should enable their detection. Specifically, when the pulsar radio signals travel through an isotropic unpolarized stochastic background of quadrupole gravitational radiation, the correlation χ between the signals which arrive from two different directions separated by angle θ may be calculated to produce the *Hellings and Downs* curve, calculated by R. Hellings and G. Downs in 1983. For two Earth-pulsar lines of sight at an angle θ from each other, the curve is

$$\chi(\theta) = \frac{1}{2}\left[1 - \frac{1}{2}\sin^2\frac{\theta}{2} + 3\left(\sin^2\frac{\theta}{2}\right)\ln\left(\sin^2\frac{\theta}{2}\right)\right]. \tag{27.1}$$

CHAPTER 28

Polarization of Cosmic Microwave Background

28.1 "FIRST LIGHT"

In late March 2014, a news item from the world of astrophysics appeared in mainstream media. "Space Ripples Reveal Big Bang's Smoking Gun" announced the captions. These news relied on the results of an ambitious experiment BICEP2. Unfortunately, the announced discovery did not get confirmed by the followup analysis. However, the method of detecting primordial GWs through the analysis of *Cosmic Microwave Background Radiation* (CMBR), used in the 2014 announcement is sound and may yet come to fruition.

This method relies on the study of the polarization of the "first light." The "first light" is, actually, not light, but microwave EM radiation which last interacted with matter just as it was transitioning from ionized plasma to neutral gas. This radiation, the CMBR, permeates the entire universe, and is observed in all direction. This is a thermal radiation, its frequency spectrum is the spectrum of a blackbody at the temperature 2.72 Kelvin (Fig. 28.1).

Although the temperature of this radiation is uniform, the minute variations on the temperature have been measured and carefully mapped out by numerous experiments in increasingly great details.

This radiation is the remnant of an earlier epoch in the evolution of the universe. The universe formed in an extremely dense and hot state. For the first 380,000 years its temperature was so high that nuclei could not form, and only plasma of elementary particles existed. The plasma behaved as a relativistic electrically charged fluid. It supported internal currents and acoustic vibrations. The photons were not free to propagate within the plasma, but constantly interacted and were scattered by it.

As a result of the expansion, the universe cooled to the temperature (approximaterly 3000 K) where protons could capture electrons and form light nuclei (primarily hydrogen). This era is called *recombination* (this name is a misnomer, since atoms had never existed until then).

Once recombination occurred, the plasma disappeared, and the interaction of the photons with the plasma stopped. Those photons now constitute the cosmic microwave background. Since the moment of recombination the photons continue to propagate freely through the vacuum. The only continuing influence on these photons is the expansion of the universe, which uniformly *redshifts* them (that is, uniformly increases their wavelength).

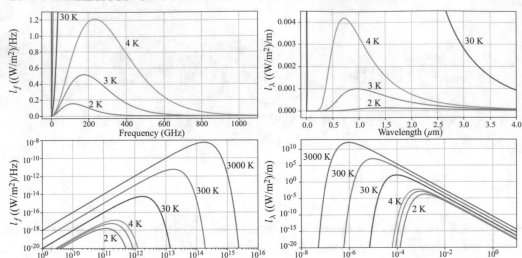

Figure 28.1: The blackbody spectrum. The curves are shown both using linear axes, as is usually done in the introductory physics books, and using log-log axes.

The salient point is that the frequency and polarization of this radiation is a "snapshot" of the state of the plasma at the last moment it existed, 380,000 years after the Big Bang. The variations in the temperature of the CMBR reflect the area where the plasma at recombination was more or less dense. These variations, on the order of 200 μK, have been carefully mapped by Wilkinson Microwave Anisotropy Probe spacecraft which operated between 2001 and 2010, and are continually refined by the Plank spacecraft launched in 2009.

28.2   POLARIZATION OF EM RADIATION

The presence of primordial GWs cannot be deduced from the measurements of the temperature variations. Instead, it is necessary to measure the *polarization* of the CMBR. The polarization of an EM wave is determined by the behavior of the electric field vector. This vector always remains in the plane perpendicular to the direction of propagation of the EM wave. Within that plane it can rotate while also changing its magnitude. As a result, the end of the vector of electric field describes an ellipse.

Such an ellipse is commonly described mathematically using *Stokes vectors*. These are four-component vectors:

$$S = \begin{bmatrix} S_0 \\ S_1 \\ S_2 \\ S_3 \end{bmatrix} = \begin{bmatrix} I \\ Q \\ U \\ V \end{bmatrix}. \tag{28.1}$$

Figure 28.2: All sky map of the CMBR.

The first entry, I, describes the light intensity. The next entry, Q, describes the degree to which the polarization ellipse is stretched along the vertical (in which case Q is a positive number) or along the vertical (Q is negative) directions. Similarly, U describes the degree to which the polarization ellipse is stretched at the $+45°$ (positive U) or at $-45°$ (negative U). Finally, the last entry V describes whether the degree of circular polarization, with the sign determining right- versus left-handedness of elliptical polarization.

28.3   THOMSON SCATTERING

The CMB originated from the last *Thomson scattering* of photons on the electron before the latter got captured by protons to form atoms. This type of scattering was described, originally, classically by J. J. Thomson.

Suppose that the plane-polarized EM wave moving along the $\hat{z}$-axis is incident upon a free electron at the origin, as shown in Fig. 28.3. The EM wave induces the electron to oscillate in simple harmonic motion. If the incident wave was linearly polarized in the $\hat{x}$-direction, the electron oscillates along the $\hat{x}$-axis. Oscillations of the electron along the $\hat{x}$-axis cannot produce re-radiation along the same $\hat{x}$-axis. On the other hand, if the incident EM wave was linearly polarized in the $\hat{y}$-direction, there is re-radiation along the $\hat{x}$-axis. Thus the re-radiated EM wave moving in the $\hat{x}$-direction has linear y-polarization.

Thus, in Thomson scattering, at the right angle to the direction of the incident radiation, the re-radiated EM wave is plane-polarized. This effect, however, "washes out" in a uniformly

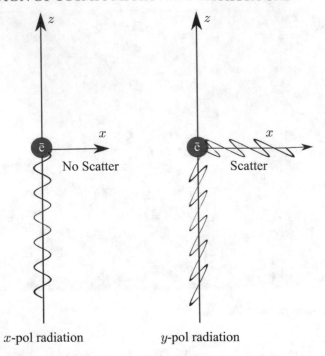

Figure 28.3: Polarization produced by Thomson scattering. (a) When an electron is illuminated by an x-polarized EM wave, the electron is induced to oscillate along the x-direction. An electron which oscillates along the x-direction cannot radiate in the same x-direction. Thus, no radiation is scattered along x-direction. The electron radiates in the y- and z-directions, but that scattered radiation is not shown in the picture. (b) When an electron is illuminated by a y-polarized EM wave, the electron is induced to oscillate along the y-direction. It then radiates in the x- and z-direction. Thus, there is scatter in the x-direction, which is shown.

dense plasma. In this case the photons are incident on the electrons from *all* directions (rather than just from the $\hat{z}$-direction). If the incoming radiation is fully isotropic, then the outgoing scattered radiation remains unpolarized. In the above example, the outgoing radiation going in the $\hat{x}$-direction has both y and z polarization components: the radiation incident from the $\hat{z}$-axis produced the outgoing y-polarization, and the radiation incident from the $\hat{y}$-axis produced the outgoing z-polarization.

28.4   PLASMA FLUCTUATIONS PRODUCE POLARIZATION

If the plasma is not uniform, however, then the amount of radiation an electron receives from the directions along the $\hat{z}$- and $\hat{y}$-axes may be different. In particular, if the temperature of the plasma varies at 90°, then the number of photons incident on an electron from, say, the $\hat{x}$-direction is different from the number of photons incident on the same electron from the $\hat{y}$-direction. The scattered radiation then, again, becomes polarized. Such temperature variations are called *quadrupole anisotropy*.

28.5   SOURCE OF PLASMA FLUCTIATIONS

In the previous chapters the word "quadrupole" has been associated with the type of radiation from localized sources. In the context of CMB polarization this word refers to an entirely different circumstance. The quadrupole anisotropy is so named not because it was produced in some sort of multipole EM or gravitational radiation. Rather, it refers to a specific pattern of temperature distributions: a distribution such, that from the point of view of some "observer electron" the temperature around it varies between high and low at 90° turns.

In every situation when the quadrupole anisotropy was present, the CMB is linearly polarized. There are no other polarization (circular or elliptical) that are observed in the CMB.

The quadrupole anisotropy in temperature of primordial plasma could have been created by a variety of mechanisms. One mechanism is the density fluctuation in the plasma. Another mechanism involves rotational motion of the parts of the plasma. Finally, the third mechanism is due to the GW.

These different mechanisms led to different perturbations of the primordial plasma. The density perturbations are called *scalar perturbations*, the perturbations due to the rotations are the *vector perturbations*, and the GWs caused *tensor perturbations*.

Even though each of these types of perturbations led to quadrupole anisotropy, in each case the result is the *linear polarization* of the resulting CMB.

The remarkable fact is that even though it is only linear polarization that we observe, the different causes of perturbations can be determined. The observable difference comes from *the distribution of the linear polarization*.

The different patterns of distribution of linear polarization are categorized into E-modes and B-modes (Fig. 28.4). These modes do not originate from the differences in the electric or magnetic fields. Neither do they have any relationship to the gravielectric and gravimagnetic fields discussed in the earlier chapters. These are entirely separate names, given only to the patterns of distribution of the linear polarization of the CMBR.

Although all three types of perturbations of the primordial plasma can give rise to both the E and the B-modes of distribution, the B-modes predominate in the perturbations caused by tensor perturbations. Therefore the discovery of such patterns in the polarization of CMB would indicate the presence of the primordial GWs at the time of recombination. The BICEP-2 experiment did discover such patterns (Fig. 28.5). However the follow-up analysis demonstrated that these originated not from the originally emitted light, but were imprinted on the CMB by the cosmic dust in the billions years afterward.

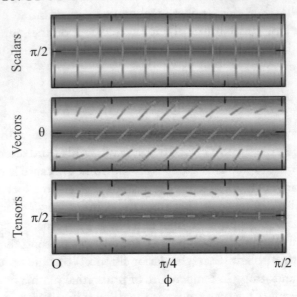

Figure 28.4: Different patterns in the distribution of the linear polarization in the CMB. Adopted from [33].

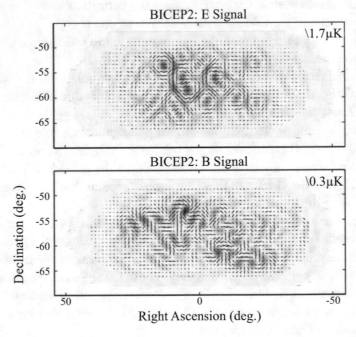

Figure 28.5: The originally reported results of BICEP2 experiment.

Bibliography

[1] Wolf, A., Van Hook, J., and Weeks, E., Electric field line diagrams don't work, *American Journal of Physics*, 64(715):714–724, 1996. DOI: 10.1119/1.18237. 12

[2] Griffith, D. J., *Introduction to Electrodynamics*, 4th ed., Pearson Education, 2015. DOI: 10.1119/1.12574. xiii, 38, 48

[3] Feynman, R. P., Leighton, R., and Sands, M., *Feynman's Lectures on Physics*. 38

[4] Marion, J. B. and Thornton, S. T., *Classical Dynamics of Particles and Systems*, 5th ed., Thomson, Brooks/Cole, 2012. DOI: 10.1063/1.3048143. xiii, 31

[5] Goldstein, H., Poole, C., and Safko, J., *Classical Mechanics*, 3rd ed., Pearson Education, 2011. 30

[6] Purcell, E. M., *Electricity and Magnetism*, 3rd ed., Cambridge University Press, 1980. DOI: 10.1017/cbo9781139005043.

[7] Pais, A., *Subtle is the Lord: The Science and the Life of Albert Einstein*, Oxford University Press, 2005. DOI: 10.1119/1.13801. 65, 66

[8] Hartle, J. B., *Gravity: An Introduction to Einstein's General Relativity*, 5th ed., Pearson Education, 2003. xiii

[9] Rindler, W., *Relativity: Special, General, and Cosmological*, 2nd ed., Oxford University Press, 2006. DOI: 10.1063/1.3021868. 66

[10] Satterly, J., The moments of inertia of some polyhedra, *The Mathematical Gazette*, 42(339):11–13, 1958. DOI: 10.2307/3608345.

[11] Lorimer, D. R. and Kramer, M., *Handbook of Pulsar Astronomy (Cambridge Observing Handbooks for Research Astronomers)*, Cambridge University Press, 2004.

[12] Bedford, D. and Krumm, P., On relativistic gravitation, *American Journal of Physics*, 52(889):889–890, 1985. DOI: 10.1119/1.14358.

[13] Jackson, J. D., From Lorenz to Coulomb and other explicit gauge transformations, *American Journal of Physics*, 70(917):917–928, 2002. DOI: 10.1119/1.1491265. 49

[14] Schutz, B. F., Gravitational waves on the back of an envelope, *American Journal of Physics*, 52(5):412–505, 1985. DOI: 10.1119/1.13627.

[15] Campbell, W. B. and Morgan, T. A., Maxwell form of the linear theory of gravitation, *American Journal of Physics*, 44(10):356–365, 1976. DOI: 10.1119/1.10195.

[16] Campbell, W. B. and Morgan, T. A., Gravitational radiation: Sources, *American Journal of Physics*, 44(11):1110–1115, 1976. DOI: 10.1119/1.10572.

[17] Nichols, D., et al., Visualizing spacetime curvature via frame-drag vortexes and tidal tendexes: General theory and weak-gravity applications, *Physical Review D*, 84(124014):1–25, 2011. DOI: 10.1103/physrevd.84.124014.

[18] Riles, K., Gravitational waves: Sources, detectors and searches, *Progress in Particle and Nuclear Physics*, 68:1–79, 2013. DOI: 10.1016/j.ppnp.2012.08.001.

[19] Jenet, F. A. and Romano, J. D., Understanding the gravitational-wave Hellings and Downs curve for pulsar timing arrays in terms of sound and electromagnetic waves, *American Journal of Physics*, 83(7):635–645, 2015. DOI: 10.1119/1.4916358.

[20] Everitt, C. W. F., et al., Gravity probe B: Final results of a space experiment to test general relativity, *Physical Review Letters*, 106(22):221101, 2011. DOI: 10.1103/physrevlett.106.221101. 60

[21] Baade, W. and Zwicky F., On supernovae, *Physical Review Letters*, 45(22):138, 1934. 83

[22] Hewish, A., Bell, S. J., Pikington, D. H., Scott, P. F., and Collins R. A., Observation of a rapidly pulsating radio source, *Nature*, 217:709, 1968. DOI: 10.1007/978-1-4899-6387-1_1. 83

[23] Oppenheimer, J. R. and Snyder, H., On continued gravitational contractions, *Physical Review*, 56:455–459, 1939. DOI: 10.1103/physrev.56.455. 80

[24] Hulse, R. A. and Taylor, J. H., Discovery of a pulsar in a binary system, *The Astrophysical Journal*, 195(L51), 1975. DOI: 10.1086/181708. 84, 105

[25] Weisberg, J., Nice, D., and Taylor, J., Timing measurements of the relativistic binary pulsar PSR B1913+16, *The Astrophysical Journal*, 722(2):1030, arXiv:1011.0718v1, 2010. DOI: 10.1088/0004-637x/722/2/1030. 105

[26] Einstein, A., *Investigations on the Theory of Browninan Motion*, Dover Publications, Inc., 1955. Original publication Einstein, A., on the movement of small particles suspended in stationary liquids required by the molecular-kinetic theory of heat, *Annalen der Physik*, 17:549–560, 1905. 119

[27] Smoluchowski, M., Zur kinetischen theorie der brownschen molekularbewegung und der suspensionen, *Annalen der Physik*, 21(14):756–780, 1906. DOI: 10.1002/andp.19063261405. 119

[28] Callen, H. B. and Welton, T. A., Irreversibility and generalized noise, *Physical Review*, 83(34):702–780, 1951. DOI: 10.1103/physrev.83.34. 119

[29] Hecht, E., *Optics*, 5th ed., Pearson Education, 2017. 131

[30] Abbott, B. P. et al., Observation of Gravitational Waves from a Binary Black Hole Merger, Physical Review Letters, 116(061102), 2016. 135

[31] Amato, J. C., Flying in formation: The orbital dynamics of LISA's three spacecraft, *American Journal of Physics*, 87(1):18–23, 2019. DOI: 10.1119/1.5075722. 137

[32] Norcia, M. A., Cline, J. R. K., and Thompson, J. K., Role of atoms in atomic gravitational-wave detectors, *Physical Review A*, 96:042118, 2017. DOI: 10.1103/physreva.96.042118. 144, 145

[33] Hu, W. and White, M., A CMB polarization primer, *astro-ph/9706147*. DOI: 10.1016/s1384-1076(97)00022-5. 154

[34] Paik, H. J., Superconducting tunable-diaphragm transducer for sensitive acceleration measurements, *Jounal of Applied Physics*, 47:1168, 1976. DOI: 10.1063/1.322699. 124

Author's Biography

DAVID M. FELDBAUM

David M. Feldbaum received his Ph.D. in experimental physics from the University of Michigan in 2003, working with trapped atoms. As a postdoctoral researcher, he spent several years at Los Alamos National Laboratory working on trapping radioactive atoms, before joining the Laser Interferometer Gravitational Observatory in Louisiana as a part of the University of Florida research group. He joined the faculty of the Southeastern Louisiana University in 2014.

Printed in the United States
by Baker & Taylor Publisher Services